FIELD MANUAL
NO. 1-100

HEADQUARTERS
DEPARTMENT OF THE ARMY
Washington, DC, 21 February 1997

ARMY AVIATION OPERATIONS

CONTENTS

DISTRIBUTION RESTRICTION: Approved for public release; distribution is unlimited.

*This publication supersedes FM 1–100, 28 February 1989.

Section II. ARMY AVIATION MISSIONS

Section III. AVIATION OPERATIONS

PREFACE

Field Manual (FM) 1–100 is Army aviation's capstone manual. It embodies the doctrinal tenets for the employment of aviation and establishes the basis for understanding aviation as an essential element of combat power. Doctrinally describing Army aviation, the manual serves as the foundation that captures the essence of Army aviation and those principles upon which it is employed. It prescribes doctrine above the level of tactics, techniques, and procedures.

This field manual provides general guidance concerning the employment of Army aviation on the modern battlefield. It forms the basis for understanding Army aviation's unique contribution to maneuver, combat support, and combat service support missions. It establishes the principles of Army aviation employment for warfighting and other operations, and applies to all echelons of aviation operations.

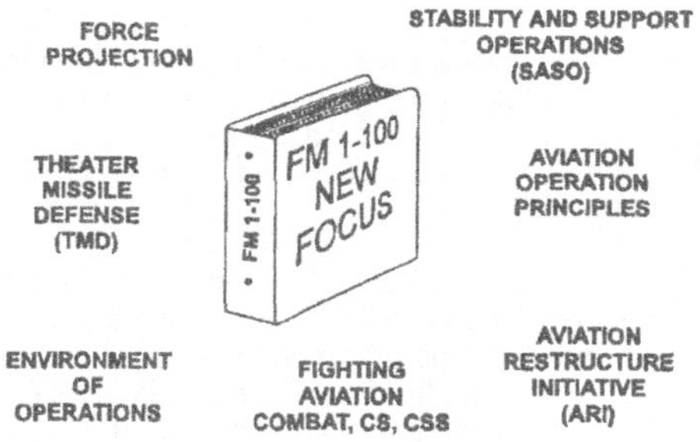

This manual is intended for use by Aviation commanders, staff officers, and all soldiers within aviation units and theater, corps, division, and brigade commanders and their staffs. It applies to all members of the combined arms team; joint, multinational, and interagency operations; and special operations, or contingency, forces that operate with Army aviation forces.

The proponent of this publication is HQ TRADOC. Submit changes to improve this manual on DA Form 2028 (Recommended Changes to Publications and Blank Forms); forward it through the aviation unit commander to the Commander, US Army Aviation Center and Fort Rucker, ATTN: ATZQ–TDS–D, Fort Rucker, AL 36362–5000.

Unless this publication states otherwise, masculine nouns and pronouns do not refer exclusively to men.

This publication has been reviewed for operations security considerations.

FUTURE DOCTRINE

This edition of FM 1–100 is written to carry Army aviation forward to the turn of the century. Our vision lies beyond the turn of the century, however. As this manual is being drafted and staffed, we are concurrently developing the new concepts that will evolve into the doctrinal foundation for the *next century.* This is a dynamic period of innovation and change.

Our leap–ahead reconnaissance and attack aircraft—the RAH-66 Comanche and AH-64 Longbow Apache—are realities. We know their current capabilities, and can envision the future potential they bring to the *future battlefield.* We are developing future doctrine based on those capabilities. At the same time, the research and development community and industry continue to create the enabling technologies-the digital communications and other linkages—we need to fight these systems to their fullest potential. The *future battlespace* will be fluid, high tempo, and nonlinear. The traditional battlefield framework of deep, close, and rear operations will become increasingly convoluted and ambiguous.

To maintain continuity with other capstone Army doctrine, this edition of FM 1–100 will continue to refer to close, deep, and rear operations; however, in the near future, a more viable framework may be simply close and extended operations. On the *nonlinear/non-contiguous battlefield of the future,* we must be prepared to conduct seamless, simultaneous operations in all directions.

Army operations will be conducted in the context of an ever-changing world. No longer can we model the force and develop our doctrine against one known threat, or even counter the capabilities of a number of known potential adversaries. Instead, we must develop and retain the warfighting capability to win decisively across the spectrum of operations with minimum friendly casualties. This is domination–based warfare-massing not only our forces, but total lethal and nonlethal effects, throughout the battlespace to dominate any potential adversary. *Army aviation is uniquely suited to these challenges.*

The Apache and Comanche fully exemplify the inextricable linkage between maneuver and fires. With unmanned aerial vehicles (UAVs) to extend their range and coverage-digitally cued by the Joint Surveillance Target Attack Radar System (JSTARS), Army airborne command and control system (A^2C^2S) UH–60 Black Hawks, and other ground–based command posts—these aircraft provide commanders with real–time intelligence and situational awareness. They maneuver throughout the depth of the battlespace to deliver precision fires with devastating lethality.

Shaping Army aviation for future operations is more than merely delivering lethal fires ...it is more than killing enemy tanks and artillery.. .it is, instead, creating a new synergy—a total integration into what is termed a *pattern of operations.* (The subsets of the pattern are depicted in italics to indicate that they are *emerging doctrinal* terms.)

Our future Aviation units will be modular and deployable. They will provide joint force commanders with a lethal and flexible force to rapidly deploy from the continental United States (CONUS), or abroad, to any theater. Deployment will be by strategic air or sea lift, self-deployment, with a maritime force aboard aircraft carriers, or by any combination of those means.

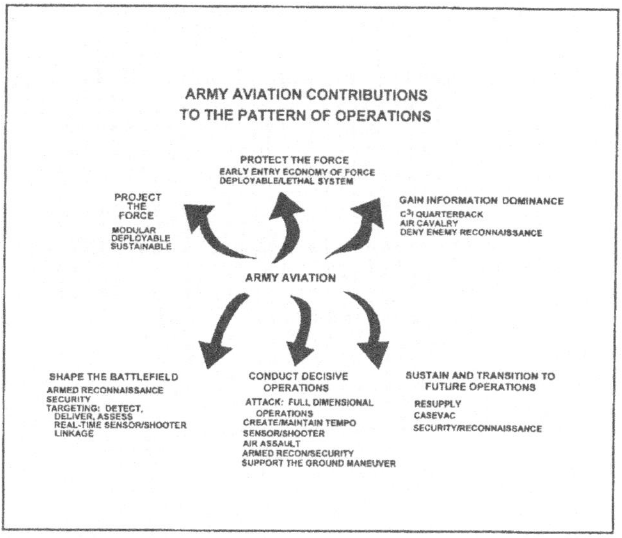

No other force can match Army aviation's ability to rapidly *project the force* and build combat power in an immature theater. Once on the ground, we become the principal means to *protect the force* as the other ground forces continue to deploy and flow into the initial lodgement. This is best exemplified by the initial days and weeks of Desert Shield as aviation units quickly deployed to Saudi Arabia and became the principal combat power for the initial covering force.

Throughout the future fight, Army aviation will be at the forefront of *gaining information dominance*. The Comanche and Longbow Apache, coupled with UAVs and the A^2C^2S UH–60, form a team that becomes, in effect, the command, control, communications, and intelligence (C^3I) key facilitator for the *future battlefield*. We can eliminate the enemy's reconnaissance, attack his command and control (C^2), and gather intelligence, while providing security for our own intelligence and C^2 systems. Digitally cued by JSTARS and other airborne and ground sensors, our future aircraft will add a new dimension of precision economy of force. Concurrently, these missions also contribute to Army aviation's key role in *shaping the battlespace*.

By conducting armed reconnaissance and security missions with real–time, sensor–to-shooter linkages, Army aviation can rapidly confirm the enemy's intentions, disrupt his tempo, deny his freedom of action, and get into his decision cycle. The ultimate in shaping the battlespace is to preclude the necessity for conducting decisive operations. We can sustain the tempo of the fight, attacking with depth and simultaneity throughout the battlespace. At a time and place of our choosing, we will initiate decisive operations in conjunction with maneuver ground forces to complete the destruction or defeat of enemy forces.

We will sustain the force and transition to future operations with combat support and combat service support provided by our UH–60 Black Hawk and CH–47 Chinook aircraft, and by air assaulting forward-operating bases from which follow–on combat operations can be conducted. We will also continue to provide the reconnaissance, security, and attack helicopter support to sustain the fight and protect the force as we prepare for follow–on operations.

Army aviation must adapt quickly to the inevitable changes that affect our mission. Our doctrine, tactics, techniques, and procedures must reflect those changes and be responsive to the needs of our units in the field. We encourage your comments and ideas as we develop our collective vision for shaping the future of Army aviation.

CHAPTER 1

INTRODUCTION TO ARMY AVIATION

There is "the enduring reality of the unknown and the uncertain; not just across the Atlantic and Pacific, but in all regions of the world that continue to harbor danger and turmoil; regions where crisis will occur when least expected." To meet this reality, contingency forces "provide global crisis and contingency response capability across the spectrum of conflict from counterinsurgency to major conventional conflict."

General Colin Powell
A Critical Analysis of the Gulf War

1-1. PURPOSE

The purpose of our capstone doctrine is to capture the essence of Army aviation and those principles upon which it is employed across the range of military operations.

1-2. STRATEGIC REALITIES

a. Recent events have underscored the uncertainty of these times. The post Cold War period has placed unprecedented operational demands on the Army. Civil disturbances, disaster relief, humanitarian and peacekeeping operations, and the threat of lesser regional contingencies punctuate the need for a trained and ready contingency-oriented Army. Amidst these global demands, domestic change and fiscal constraints broaden the challenge.

b. This era also confirms the application of high technology in future warfare. Weapons with the *"effects of massed forces"* are available to any nation possessing hard currency. Precision munitions, digital communications, and position location equipment promise to change the face of future battle.

c. The physical and intellectual dimensions of battlespace urgently demand intuitive and versatile leaders supported by agile battle staffs and well–trained soldiers. Mobility, agility, simultaneity of effort, lethality, increased battle tempo, and space-age logistics must dominate the Army's restructuring initiatives and investment decisions.

1-3. ARMY'S RESPONSE

a. The Army has responded to this new environment with continental United States (CONUS)–based contingency and reinforcing forces and some forward-deployed units. Total Force initiatives are underway among the Active and Reserve Components to give broadened meaning to the doctrine development of a trained and ready Total Army, capable of decisive victory. Force restructuring initiatives are being implemented to leverage high technology for a downsized force. Modernization decisions are focused on projecting and sustaining the force, protecting deployed forces, winning the information war, conducting precision strikes, and dominating the maneuver battle. The result is a combined arms team that leverages all dimensions of the ground regime.

b. Aviation, as a maneuver force, is the third dimension centerpiece of the land force. Reconnaissance, attack, utility, and cargo helicopters complemented by special operations forces (SOF), fixed-wing and medical evacuation (MEDEVAC) aircraft, and air traffic service (ATS) units, comprise our contribution to the fight for a global Army. While the range of military operations demands readiness for a wide range of employment, warfighting is our mission and we cannot lose sight of this obligation.

1-4. A VISION

a. As we look toward the next century and the pivotal role of Army aviation across the full range of military operations, it is imperative that we have a vision—a concept that will serve to guide our collective thought and actions—as we look to the future (Figure 1-1).

b. Although we emphasize and have soundly demonstrated our versatility and proficiency in stability and support operations (SASO), Army aviation's primary focus remains with combat operations. That focus on warfighting is guided by immutable principles that have stood the test of time and the trials of war.

AVIATION VISION

AVIATION IS THE RELEVANT FORCE FOR THE 21ST CENTURY PROVIDING COMBAT, COMBAT SUPPORT, AND COMBAT SERVICE SUPPORT CAPABILITIES ACROSS THE SPECTRUM OF FULL-DIMENSIONAL OPERATIONS. ITS INHERENT VERSATILITY, MANEUVER ADVANTAGE, AND WARFIGHTING EFFECTIVENESS WILL INFLUENCE ALL DIMENSIONS OF THE FUTURE BATTLESPACE. HIGHLY MOTIVATED AVIATION SOLDIERS, EQUIPPED WITH MODERN SYSTEMS AND TRAINED TO WORLD CLASS PROFICIENCY, WILL PROVIDE COMMANDERS AT ALL LEVELS AN EXPONENTIAL INCREASE IN LETHALITY, THE LEADERSHIP TO HARNESS THE TECHNOLOGICAL REVOLUTION OF THE DIGITAL BATTLEFIELD, AND THE ABILITY TO ACHIEVE DECISIVE VICTORY.

Figure 1-1

1-5. AVIATION OPERATIONAL PRINCIPLES

Mission planning and execution are driven by general principles that apply and go beyond the principles of war and the tenets of Army operations. These general principles are as follows:

a. Aviation operates in the ground regime.

(1) This cardinal principle defines aviation's role as an element of landpower. Aviation is a component of the combined arms team, not the air component of the US Army.

(2) Aviation's primary mission is to fight the land battle and to support ground operations. Aviation is comprised of soldiers, not airmen, and its battlefield leverage is achieved through a combination of reconnaissance, mobility, and firepower that is unprecedented in land warfare.

(3) Aviation greatly enhances the commander's ability to apply four fundamental principles of war—maneuver, mass, surprise, and economy of force.

b. Aviation expands the battlefield in space and time at each echelon.

(1) Expansion of the battlefield is necessary to enable the commander to seize the initiative at a critical point in the battle. Aviation expands the ground commander's battlefield, principally in space and time, by extending the range at which direct fires and observed fires can be concentrated on the enemy; and by expanding his reconnaissance and surveillance envelope beyond the effective range of other systems.

(2) Aviation expands battlespace at each echelon to which it is assigned or attached—providing a capability where none previously existed or enhancing existing capabilities. Aviation allows commanders to achieve the effects of mass without massing weapons systems.

c. Aviation performs combat, combat support (CS), and combat service support (CSS) battlefield functions (Figure 1-2).

(1) Aviation's greatest contribution to battlefield success is the ability it gives the commander to apply decisive combat power at critical times, virtually anywhere on the battlefield. This may be direct fire from aviation maneuver units or the insertion of overwhelming infantry forces or artillery fires, delivered into combat via air assault. This versatility is the very essence of Army aviation.

(2) CS missions support ground combat operations. These operations include air movement; command and control (C^2); ATS; electronic warfare; close in fire support; support by fire; combat search and rescue; and aerial mine delivery. The primary function of these missions is to support combat elements in contact with the enemy.

(3) Aviation performs CSS functions in support of units throughout the entire area of operations. Aviation units enhance the commander's battlespace through rapid delivery of supplies and personnel and aeromedical evacuation.

d. Aviation is concentrated at division and corps level.

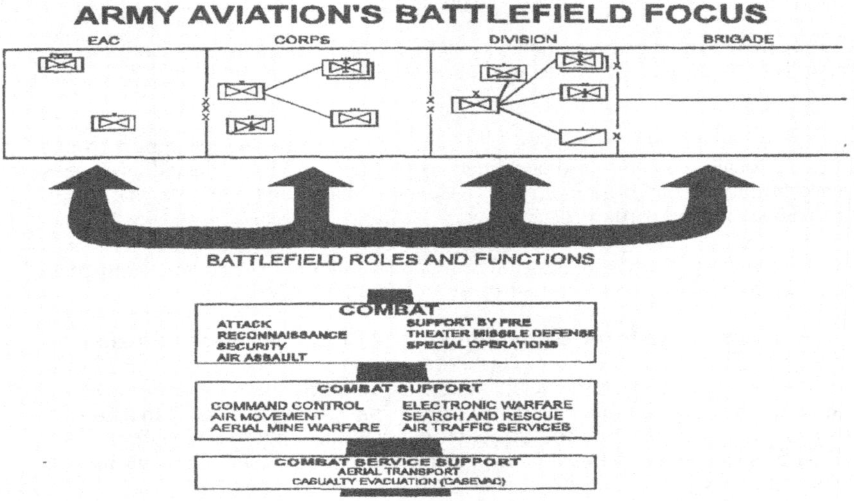

Figure 1-2

(1) The corps aviation brigade may operate directly for the corps commander or be placed under operational control (OPCON) of a subordinate division. The corps commander can task organize other corps assets, especially division aviation units, under the command of the corps aviation brigade or task the corps aviation brigade to support an armored cavalry regiment (ACR).

(2) The aviation brigade may also be tasked to be a covering force headquarters when augmented by ground forces. The corps aviation brigade conducts attack and reconnaissance operations to find, fix, and destroy enemy forces; it also conducts security, air assault, C^2, and air movement operations throughout the corps area of operations (AO).

(3) The corps aviation brigade plans, coordinates, and executes aviation operations in support of the corps scheme of maneuver. It can be expected to operate anywhere in the corps area.

(4) The division aviation brigade conducts all aviation combat, CS, and CSS missions (except ATS and fixed-wing operations) in support of the division scheme of maneuver. The primary mission of the division aviation brigade is to find, fix, and destroy enemy forces within the division area. The division aviation brigade can accomplish this mission as an aviation–pure or task–organized force.

(5) Combined arms battles and engagements are fought by brigades and divisions. Division is the lowest level at which all of the combined arms are normally integrated.

(6) The combination of infantry, armor, and aviation is a habitual association at the division level. All three arms are required for operations, in depth, throughout the course of battle. Therefore, combat aviation must be primarily assigned to, and employed by, divisional aviation brigades, just as infantry and armor battalions are assigned to, and employed by, their parent brigades.

(7) Aviation forces fight as units and must be given unit missions. Aviation units conducting tactical operations are given maneuver objectives rather than individual targets.

e. Aviation units are integrated into the combined arms down to the level at which they will be employed.

(1) The division aviation brigade is the primary level of integration. The brigade commander is responsible for the operation of all divisional aviation; he will normally command and integrate additional aviation units attached or under OPCON from corps.

(2) When aviation units are placed under OPCON of the other maneuver brigades, they normally will be on a mission basis and tailored or task organized with assets from brigade and/or division. A liaison detachment should be placed at the ground brigade command post to improve synchronization and responsiveness, especially in changing tactical environments.

f. Planning times for aviation and ground maneuver elements will be the same.

(1) Aviation units conduct deliberate planning within the same time parameters as the other maneuver elements. Airspace coordination, route clearances, and weather updates complicate the task for aviation staffs; however, for effective combat operations, the standard is the same.

(2) Both ground and air mission planning times can be reduced when plans are carefully integrated, effective liaison occurs, and standing operating procedures (SOPs) are optimized.

1-6. BATTLEFIELD OPERATING SYSTEM

The battlefield operating system (BOS) is comprised of the major functions performed on the battlefield. These functions facilitate the integration, coordination, preparation, and execution of successful combined–arms operations to successfully execute Army operations (battles and engagements) and accomplish military objectives directed by the operational commander. They include *intelligence; maneuver; fire support; mobility, countermobility,* and *survivability; air defense; logistics;* and *battle command* (Figure 1-3).

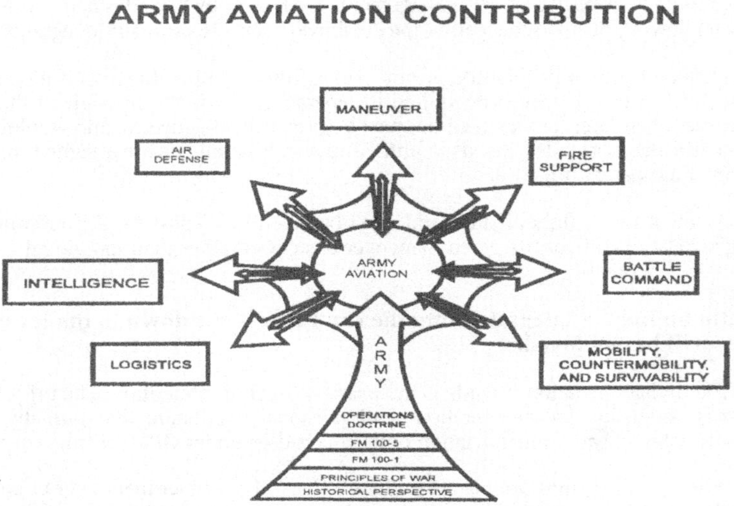

Figure 1-3

Commanders use the BOS to integrate and coordinate these functions to synchronize battle effects in time, space, and purpose. Army aviation contributes to all BOS functions addressed in the following paragraphs:

a. ***Maneuver.***

(1) Maneuver is defined as *"Employment of forces on the battlefield through movement in combination with fire, or fire potential, to achieve a position of advantage in respect to the enemy in order to accomplish the mission."*

(2) During decisive operations, Army aviation's mobility and firepower make it a dominant force—a force that can gain and maintain contact; destroy the enemy in depth; attack decisive points at the tactical and operational levels; and allow him no safe haven in which to reorganize, rearm, or recover.

(3) Attack helicopter units give the commander a force that can rapidly build devastating firepower at any point on the battlefield.

(4) Army attack helicopters can support the close fight by securing an armored or mechanized force's flanks-providing aerial fires, target acquisition, and reconnaissance. They can also attack decisive points and critical targets hundreds of kilometers (km) deep in the enemy's rear area simultaneously.

(5) By destroying follow-on forces, C^2 nodes, and logistical supply assets before they can be employed against friendly forces, aviation can significantly influence tomorrow's close fight.

(6) Deep operations require precise synchronization of both lethal and nonlethal assets; aviation performs not only maneuver, but supports other maneuver forces with fires and maneuver. Since this is true, aviation commanders are accustomed to massing effects on the battlefield. We can rapidly mass effects; then just as rapidly shift our focus to a new main effort. This flexibility and versatility are paramount to decisive operations.

(7) UH–60 Black Hawk and CH–47 Chinook units also play a pivotal role in combat operations. The means to project a forward-operating base across hundreds of kilometers allow the friendly force commander to define the battlespace, control it, and engage the enemy at a time and place of his choosing.

(8) UH–60 Black Hawk and CH–47 Chinook units can rapidly move dismounted troops, artillery, and antitank weapons anywhere on the battlefield to attack targets; seize critical terrain; or cut off an enemy's retreat so he can be destroyed in place.

b. ***Intelligence.***

(1) Intelligence is the product resulting from the collection, analysis, and dissemination of all available information that is immediately or potentially significant to military planning and operations.

(2) The commander drives intelligence by specifying what his intelligence and targeting requirements are; and requiring his intelligence BOS to provide the intelligence he needs, in the format he can use, in time to support his decision–making process.

(3) The commander's priority intelligence requirements (PIRs) will drive this process. The tasks required to properly integrate intelligence into aviation missions present a challenge for aviation commanders at every level. Primary intelligence tasks are—

- Provide indications and warnings.
- Perform intelligence preparation of the battlefield.
- Perform situation development.
- Perform target development and support to targeting.
- Support force protection.
- Perform battle damage assessment (BDA).

(4) Aviation augments intelligence collection by providing reconnaissance, early warning, target acquisition, electronic support (ES), and BDA.

(5) Army aviation also assists the intelligence effort by conducting missions to attack the enemy's command, control, and intelligence (C^2I) systems; and by conducting missions to protect friendly C^2I.

(6) Army aviation provides the commander with near real–time intelligence throughout his battlespace with its attack and cavalry aircraft and special electronic mission aircraft (SEMA). In fact, with the OH–58D Kiowa Warrior and AH–64 Apache, a single combat system can find, fix, and observe or destroy enemy assets across the depth of the battlefield.

(7) Aerial exploitation battalions (AEBs) exist in most Army corps; they provide an organic deep look capability for the corps commander, focusing on second–echelon forces that can influence the fight greater than 72 hours into the battlespace. The Guardrail Common Sensor can provide targetable communications intelligence (COMINT) and electronic intelligence (ELINT) on enemy targets as far as 300 km away. Besides Guardrail, Airborne Reconnaissance Low (ARL) provides all–source imagery and signals intelligence throughout the range of military operations.

(8) At the division and armored cavalry regiment, the EH–60 Quickfix is an important SEMA asset for conducting intelligence and electronic warfare (IEW). The EH–60A (Quickfix) and the follow on EH–60L (Advanced Quickfix) provide the commander with signal intelligence and electronic jamming capability using the advantage of aviation mobility.

(9) Intelligence is critical to the successful conduct of aviation operations—particularly deep operations. Army aviation units often require joint, theater–level intelligence support; joint and echelon above corps (EAC) assets must be integrated into the aviation collection plan. This is particularly vital to engagement area (EA) planning and development. The intelligence links necessary to *"see"* an EA must be emplaced in a timely manner and continuously monitored.

(10) Another critical area that requires the same level of detailed planning and joint/EAC support is joint suppression of enemy air defense (JSEAD). JSEAD is more than planning artillery fires. It is a synchronized plan that integrates all available lethal and nonlethal joint assets into an operation concentrating on dismantling the enemy's entire air defense (AD) network—not simply isolating and suppressing or destroying specific weapons. This more thorough approach requires continuous and detailed intelligence collection and assessment.

c. ***Fire Support.***

(1) Fire support operations are conducted throughout the wide range of military operations. Fire support includes the delivery of conventional and smart munitions by armed aircraft, land– and sea–based fire systems, and electronic warfare (EW) systems against ground targets. Operations often hinge on carefully planned integration of fires.

(2) Army aviation, as a maneuver force, contributes to fire support operations by acquiring targets; providing laser designation; adjusting indirect fires; and providing command and control to artillery units. Aviation units also contribute to fire support by engaging targets with close in fire support and conducting support by fire missions.

(3) The EH-60 Quickfix mission contributes to fire support by providing *"electronic"* fires in the form of signal jamming and electronic deception. FM 100–5 states *"when developing the concept of operation, tactical commanders should consider EW assets the same as they do artillery."*

d. ***Air Defense.***

(1) Across the wide range of military operations, commanders at all echelons are faced with an increasingly capable air and missile threat. Today's widespread technological advances are challenging the maneuver commander in his execution of air and ground maneuver. All commanders can expect the enemy to violently contest the use of the airspace at any level of conflict with an extensive array of weapon systems.

(2) The air dimension of the battlefield must be effectively controlled by disrupting, degrading, or deceiving enemy air defenses. Suppression of enemy air defense (SEAD) prevents effective fires on friendly forces. Thus, Army aviation and tactical air assets can maneuver into the depth of the enemy to weaken his ability and will to fight.

(3) SEAD and JSEAD are major functional areas that affect the operations of all combined arms actions. Commanders at operational and tactical levels must coordinate and allocate a balance of resources (direct, indirect, electronic attack) to SEAD/JSEAD. Aviation commanders must be involved in recommending and developing SEAD and JSEAD priorities. As evidenced in Desert Storm, Army aviation not only may be a benefactor of SEAD/JSEAD operations, we also may be called upon to provide SEAD/JSEAD fires at the strategic, operational, and tactical levels of war.

(4) AD operations are performed by all members of the combined arms team; however, ground-based air defense artillery (ADA) units execute the bulk of the force protection mission. AD operations protect the force by preventing enemy aircraft, missiles, and remotely piloted and unmanned aerial vehicles (RPV/UAV) from locating and attacking friendly forces.

(5) Army aviation assists AD units by conducting theater missile defense (TMD) attack operations and contributing to short range air defense (SHORAD). Army aviation units conduct deep operations to attack threat missile components, such as launch platforms; command, control, communications, computers, and intelligence (C^4I) nodes; missile stock infrastructure; and UAV launch facilities.

(6) Army aviation can attack these targets when they are stationary or on the move. In certain environments, Army aviation can execute these missions without the benefit of sensor/eyes on target or a precise grid coordinate. Army aviation assets may also be called upon to intercept and destroy enemy helicopters and UAVs that pose a threat to friendly forces.

e. ***Mobility, Countermobility, and Survivability.***

(1) Mobility operations preserve the freedom of maneuver. They include breaching enemy obstacles; increasing battlefield circulation; improving existing routes, or building new ones; providing bridge and raft support for crossing rivers; and identifying routes around contaminated areas.

(2) Army aviation contributes to the mobility and survivability of the force by overcoming both man–made and natural obstacles. Aerial reconnaissance elements identify obstacles in the path of advancing forces and search for bypass routes or safe crossing sites. This precise information saves valuable time and helps the force continue to move unimpeded.

(3) Aviation forces also provide security during obstacle-emplacement or crossing operations by rapidly moving troops and supplies to secure obstacle locations or crossing sites.

(4) Countermobility missions hinder enemy maneuver. Aerial delivered mines can be employed to emplace tactical minefield; reinforce existing obstacles; close lanes, gaps, and defiles; protect flanks; and deny the enemy AD sites. Aerial delivered minefield can also be employed for flank protection of advancing forces and for operating in concert with air/ground cavalry units on flank guard or screen missions.

(5) Survivability operations protect friendly forces from the effect of enemy weapons systems and from natural occurrences. Hardening of facilities and fortification of battle positions are active survivability measures. Deception, operational security (OPSEC), and dispersion can increase survivability. Nuclear, biological, and chemical (NBC) defense measures are also key survivability operations.

f. ***Logistics.***

(1) Logistics entails the essential capabilities, functions, activities, and tasks necessary to sustain all elements of operating forces in theater at all levels of war.

(2) Aviation assists in providing basic sustainment operations for the Total Force. Aviation forces may support major maneuver forces, CS elements, or major CSS elements for the maneuver force.

(3) Aviation cargo and utility assets may perform force sustainment as well as support aviation–specific sustainment requirements. However, air movement is a relatively inefficient means to transport heavy supplies and equipment and should be reserved for the support of major operations in which air movement is essential for success or in situations where emergency resupply is vital for mission accomplishment.

g. ***Battle Command.***

(1) Battle command functions are performed through an arrangement of personnel, equipment, communications, and procedures employed by a commander to plan, direct, coordinate, and control forces and operations to accomplish a mission.

(2) Battle command is the art of battle decision making; leading; and motivating soldiers, and their organizations, into action to accomplish missions. Battle command consists of visualizing the current state and the desired end state for an operation. It includes deciding how to get from one state to the other at the least cost to the soldier.

(3) Battlefield visualization lies at the center of battle command. It is a continuous process that commences before an operation and continues through achievement of the desired conclusion to that operation.

(4) Visualization of the battlefield requires use of operational tools derived from science and technology. These operational tools provide the commander with near real–time information on the current situation. Situational awareness includes knowing the disposition of friendly forces, enemy forces, noncombatants, the environment, and the terrain.

(5) Army aviation—with its reconnaissance and security assets and SEMA platforms-can assist the force commander by providing accurate information in virtually all environmental conditions and throughout the full spectrum of conflict.

(6) Reliable communications are central to both battle command and battle control. Effective battle C^2 requires reliable signal support systems to enable the commander to conduct operations at various tempos. Army aviation has the capability to provide highly mobile C^2 command posts to commanders at the brigade, division, corps, and EAC levels. The communications suites in these C^2 aircraft are compatible with the force's command post mission.

(7) In addition, by using its ATS assets, aviation supports the A^2C^2 mission; it aids in the regulation, integration, and deconfliction of the flights of both Army aircraft and Joint Service aircraft as well as UAV.

1-7. TRAINING AND READINESS CHALLENGES

a. Global realities require that Army aviation be prepared for employment throughout the entire range of military operations. Several factors present unique challenges to commanders concerning the conduct of training and readiness:

(1) Long overseas deployments on short notice will be the standard.

(2) Threat forces will probably outnumber early deploring US forces and may have technological parity in some weapons systems.

(3) Early deploying forces must be mobile, lethal, survivable, and sustainable upon arrival.

(4) Integrating Army National Guard and Army Reserve forces into operations at all levels.

(5) Maintaining readiness while undergoing major force restructuring.

(6) Harnessing increased situational awareness provided by digitization.

(7) Maintaining troop morale/equipment in spite of wide range of missions.

(8) Conducting realistic training and deployments while complying with environmental regulations.

(9) Maintaining readiness with decreased home station OPTEMPO and increased frequency of deployments.

b. Seldom, if ever, will military operations be conducted by a single service. The Army will act as part of a joint or multinational force in future operations. Complementary contributions of every component add to the effectiveness of the Total Force. Aviation possesses inherent characteristics that guarantee it will play a significant, if not unique, role throughout the range of military operations (Figure 1-4).

STATES OF THE ENVIRONMENT	GOAL	MILITARY OPERATION	EXAMPLE
WAR	Fight and Win	WAR COMBAT	Large-scale combat operations... Attack Defend
CONFLICT	Deter War and Resolve Conflict	STABILITY AND SUPPORT NONCOMBAT	Strikes and raids Peace enforcement Support to insurgency Antiterrorism Peacekeeping NEO
PEACETIME	Promote Peace	STABILITY AND SUPPORT	Counterdrug Disaster relief Civil support Peace building Nation assistance
The states of peacetime, conflict, and war could exist all at once in the theater commander's strategic environment. He can respond to requirements with a wide range of military operations. Noncombat operations might occur during war, just as some stability and support operations require combat.			

Figure 1-4

1-8. FORCE PROJECTION

a. Force projection—a key element of power projection—is the ability to rapidly alert, mobilize, deploy, and operate anywhere in the world. As with Operations Just Cause and Desert Shield, force projection operations usually start as a crisis response; may require light, armored, or special operations forces; and may be either opposed or unopposed.

b. Aviation units deploying into a theater must be prepared for both offensive and defensive operations. If the threat is minor, it may be possible to enter directly into offensive operations as in Operation Just Cause. Against a formidable opponent, it may be necessary to assume a security mission or a defensive posture while forces are sufficiently built up to ensure success in offensive operations as in Operation Desert Storm.

c. Placing combat aviation forces in the early entry phase offers the ground commander a force that can provide reconnaissance, security, and C^2 over great ranges, in depth, at night; and increase his security capability during the critical phase of force buildup.

d. The presence of armed helicopters in the initial force package may deter the threat or interrupt his decision cycle long enough for additional friendly forces to arrive. If the entry force must conduct forcible entry operations to obtain a lodgement or secure the force against an aggressive threat, attack helicopters can place powerful direct fire capability in the hands of the ground commander.

e. Assault and cargo helicopters can rapidly move personnel, equipment, and supplies across great distances rapidly expanding the AO. SEMA and other fixed-wing platforms efficiently perform a wide range of intratheater reconnaissance and passenger transport missions even further enhancing the flexibility and versatility of our force.

1-9. JOINT OPERATIONS

a. Joint operations are the integrated military activities of two or more service components—Army, Navy, Air Force, Marine Corps—of the US military.

b. US joint forces must overcome joint operational and logistical differences. Complementary contributions of every service's forces add to the effectiveness of the Total Force.

c. Army aviation forces will continue to operate as part of the Army forces to a unified command, a specified command, or as part of a subordinate joint force.

d. The aviation force commander advises the joint task force commander on the capabilities, limitations, planning, and execution of aviation operations to support the joint contingency mission.

NOTE: Joint operations does not imply that planning must occur exclusively within high echelon staffs. *Joint air attack team (JAAT)* strategy evolved through direct team-level interaction with US Air Force (USAF) pilots. Refinements in joint electronic combat tactics are occurring through direct coordination between Quickfix, at the platoon level, and the USAF squadron that conducts the airborne EW mission *"Compass Call."* This type of creative interaction between service forces should be encouraged by all commanders.

1-10. MULTINATIONAL OPERATIONS

a. Multinational operations involve diplomatic–military actions between two or more agencies, with armed forces of two or more nations to achieve the strategic end state; alliances or coalitions can be formed to carry out these actions.

b. Army aviation must be prepared to conduct multinational operations with the air, land, and naval forces of allied governments.

c. Combatant commanders face numerous challenges when planning and conducting multinational operations. Each participant brings its own unique capabilities and limitations to the operation. Commanders must not only consider cultural and language differences, but also differences in equipment, doctrine, and logistics.

d. The key to success in multinational operations is matching capabilities with missions and aggressive liaison between forces.

e. Army aviation forces will normally operate as part of the US Army component during multinational operations.

f. The aviation commander will advise the Army component or allied force commander on the capabilities, limitations, planning, and execution of aviation operations.

CHAPTER 2

FUNDAMENTALS OF ARMY AVIATION OPERATIONS

Army operations require worldwide strategic mobility. Given this requirement, Army forces must have the capability to conduct operations in any environment, under any conditions. These conditions include war, peace, and conflict. Army aviation doctrine focuses on the integration and synchronization of aviation forces within the framework of the land component commander's operational concepts.

Section I. SHAPING THE BATTLESPACE

2-1. ARMY AVIATION'S UNIQUE CAPABILITIES

a. Aviation is not a substitute for any other member of the combined arms team. Rather, it brings a degree of versatility not replicated by other members of the combined arms team and a range of unique capabilities that complement those of the other combat arms.

b. Aviation maneuvers rapidly and simultaneously in the ground commander's battlespace to bring decisive combat power to bear at the decisive points and times in the area of operations (AO). There is an inextricable linkage between maneuver and fires. *Army aviation maneuvers while leveraging organic firepower to shape the battlespace or conduct decisive operations as directed by the force commander.*

c. Aviation compresses battlespace by shortening and/or mitigating the effects of time/distance factors and terrain on maneuver. Aviation forces also reduce time requirements through speed and mobility once thorough planning is complete. Aviation's ability to operate in all dimensions of battlespace provides a degree of flexibility and agility that is unique.

d. Synchronizing aviation maneuver with ground maneuver—by enhancing reconnaissance, providing security, and conducting attacks and counterattacks—allows the friendly force commander to shape the battlespace to set the conditions for the close fight and achieve a positional advantage in both time and space by altering the enemy's tempo. Linked with deep fires, aviation maneuver offers the ground commander the capability to influence events simultaneously throughout his AO.

2-2. TENETS OF ARMY AVIATION DOCTRINE

The ability to successfully fight and/or conduct war, peace, and conflict operations depends on the correct application of the five basic tenets of Army aviation doctrine. These tenets include—

a. ***Initiative.***

(1) Initiative determines or changes the terms of battle through action. In combat operations, aviation commanders set the tempo by seizing the initiative. Commanders fight tenaciously and aggressively, never allowing the enemy to recover from the initial shock of an attack. Soldiers and systems are pushed to the limits of their endurance for as long as necessary.

(2) Retaining the initiative requires planning beyond the initial operation and anticipating key events well into the future. In stability and support operations (SASO), aviation commanders take the initiative by anticipating near–and long–term personnel, equipment, and logistical support requirements relative to the operation.

b. ***Agility.***

(1) Agility is the ability of friendly forces to act faster than the enemy. In combat operations, aviation commanders exploit the agility of their units through speed, mobility, and reaction time capabilities.

(2) Technological developments in intelligence gathering, aviation mission planning, and communications have improved situational awareness during both the pre- and post–aircraft launch phases of an operation. This situational awareness provides an edge to aviation commanders in that aviation assets can now be directed to critical places at critical times on the battlefield.

(3) Furthermore, agility allows the aviation commander to rapidly rearm and refuel, get back into the fight, and continue to attack the enemy. In SASO, aviation commanders use their assets to reach locations unreachable by other means of transportation. Aviation assets can move personnel, equipment, and supplies in large quantities and in a timely manner.

c. ***Depth.***

(1) Depth is the extension of operations in time, space, resources, and purpose. In combat operations, aviation commanders understand depth as the ability to conduct simultaneous close, deep, and rear operations. Commanders sustain the momentum by taking advantage of all available resources, and attacking the enemy simultaneously in all battlefield dimensions.

(2) With access to joint and combined arms capabilities, aviation commanders can plan for and control numerous means of simultaneous or near–simultaneous ordnance delivery on multiple targets. In SASO, depth is the capability to conduct simultaneous yet different type operations.

(3) For example, attack helicopters may be required to conduct reconnaissance throughout certain areas of an AO, while CH-47 Chinook assets are transporting life support supplies in another part of the same AO, while medical relief operations are being conducted in still another part of the same AO.

d. ***Synchronization.***

(1) Synchronization is the use of time, space, and resources to produce maximum relative combat power at the decisive time and place. In combat operations, aviation commanders understand synchronization as the planned integration and execution of combat power.

(2) Synchronization requires exact coordination among the various combat, combat support (CS), and combat service support (CSS) units in any operation. Joint and multinational asset capabilities must be considered where and when applicable. In SASO, aviation commanders must synchronize both vertically and horizontally, planning phases, alert phases, deployment, early entry operations, mission execution, and redeployment.

e. ***Versatility.***

(1) Versatility is the ability of units to meet diverse mission requirements. In combat operations, aviation commanders demonstrate versatility by their ability to shift focus, tailor aviation forces, and move from one role or mission to another rapidly and efficiently.

(2) In SASO, the aviation commander recognizes mission requirements may not mirror the mission essential task list (METL), to which the aviation unit has been trained. Non–METL-based tasks require a change in focus, rapid trainup, and execution under conditions outside the normal operating environment.

Section II. ARMY AVIATION MISSIONS

Aviation units operate in the ground regime. As a fully integrated member of the combined arms team, aviation units conduct combat, CS, and CSS operations. Aviation units operate across the entire length and breadth of the AO (close, deep, and rear), and can be expected to conduct simultaneous operations, 24 hours a day.

The key to success in planning aviation maneuver in conjunction with the ground scheme of maneuver is including the aviation commander early in the planning process. Aviation missions are received by the commander and, with his guidance, the aviation headquarters will task organize forces and plan the execution of aviation operations. It is important to note that the aviation commander's AO can be as large as the division or corps AO.

2-3. AVIATION COMBAT MISSIONS

Aviation combat missions (Figure 2-1) are performed by maneuver forces engaged in shaping the battlespace and conducting decisive combat operations by employing direct fire and standoff precision weapons in combined arms operations.

AVIATION MISSIONS

- **Combat**
  - **Reconnaissance**
  - **Security**
  - **Attack**
  - **Air Assault**
  - **Theater Missile Defense**
  - **Special Operations**
  - **Support by Fire**
- **Combat Support**
  - **Command and Control**
  - **Air Movement**
  - **Electronic Warfare**
  - **Combat Search and Rescue**
  - **Air Traffic Services**
  - **Aerial Mine Warfare**
- **Combat Service Support**
  - **Aerial Sustainment**
  - **Casualty Evacuation**

Figure 2-1

a. ***Reconnaissance.***

(1) Reconnaissance operations obtain information by visual observation or other detection methods. This information may concern the activities and resources of an enemy or potential threat, or the meteorological, hydrographic characteristics of a particular area. Reconnaissance assets must possess the ability to develop the situation, process the information, and provide it to commanders in near real time. Army aviation's most modern assets, the OH–58D Kiowa Warrior and the AH–64 Apache, give the force commander a dramatically improved 24-hour air reconnaissance capability that can better develop the situation and rapidly send information to wherever it is most needed. No longer is the primary mission of attack helicopter assets within cavalry units to protect the scouts.

(2) Air reconnaissance complements and extends the zone covered by ground reconnaissance. Successful aerial reconnaissance obtains information useful in effectively directing ground reconnaissance units. Under favorable conditions, aviation furnishes early information concerning the enemy's general disposition and movements to considerable depth beyond the forward edge of the battle area (FEBA).

(3) Army aviation not only participates in the traditional missions of route, area, and zone reconnaissance, it also significantly contributes to reconnaissance–in–force. Reconnaissance–in–force is always conducted by a large enough force to place the enemy at some risk while providing self–protection. It can be conducted by an aviation-pure force or in conjunction with ground forces. Its primary purpose is to gain information and test the enemy's strength, disposition, and reaction. It is used when the enemy is known to be operating in some strength in a given area but sufficient intelligence cannot be developed by other means.

b. ***Security.***

(1) The commander conducts security operations to provide maneuver space, reaction time, and protect the main body. Security is incorporated as part of the battlefield framework in planning all offensive or defensive operations. Although reconnaissance and security missions are associated with the corps cavalry regiment and the division cavalry squadron, attack helicopter battalions are well suited for these missions.

(2) Counterreconnaissance is an inherent task in all security operations. It is the sum of actions taken at all echelons to counter enemy reconnaissance and surveillance efforts through the depths of the AO. It is active and passive and includes combat action to destroy or defeat enemy reconnaissance elements. In the execution of counterreconnaissance, air and ground cavalry units operate either offensively or defensively using whichever tactics best accomplish the task.

(3) Surveillance is also continuous during security operations. Even during security missions that involve fighting the enemy, the aeroscouts' primary task remains gathering information. Air and ground scouts are coordinated to synchronize their complimentary capabilities.

(4) Army aviation's special electronic mission aircraft (SEMA) perform surveillance at both the operational and tactical levels of war.

c. ***Attack.***

(1) The primary purpose of attack helicopter operations is the destruction of enemy ground force at decisive points. Attack units can conduct deep operations or be used in conjunction with ground maneuver units during close battle operations. For cross-component support, Army attack helicopters, usually tasked as units, can perform a close air support (CAS) function.

(2) Attack units normally are most effective when used in mass in continuous operations on the enemy's flanks and rear. Night operations are the preference. Corps attack battalions can be used independently by the corps commander or placed under OPCON of divisions to execute massed attacks on the enemy in depth.

d. ***Air Assault.***

(1) Air assault operations are those in which air assault forces (combat, CS, and CSS)—employing the firepower, mobility, protection, and total integration of helicopter assets in their air or ground roles—maneuver on the battlefield, under the control of the air assault task force commander (AATFC), to engage and destroy forces or to seize and hold key terrain. Either the ground or air maneuver commander is designated the AATFC.

(2) Air assault operations are inherently complex, fully synchronized combat operations particularly important for light forces as they are the primary means of rapid deployment. In some cases, they are the only means of employment directly into combat. Air assault should always be considered by heavy forces to assist in overcoming obstacles in the seizure of critical terrain, and in follow and support missions to preserve the momentum of attack.

(3) The level of precision required to successfully conduct air assault operations requires deliberate planning and the detailed synchronization of all battlefield operating systems (BOSs).

(4) Air assault security is provided by air cavalry and attack units in coordination with conventional fire support to set conditions before the air assault and to continue to provide supporting fires once the air assault force is established on the ground.

e. ***Theater Missile Defense.***

(1) The theater missile threat is real and increasing in scope. Proliferation of theater ballistic missiles (TBMs) presents a serious threat to maneuver forces during many potential contingencies. While the risks from fixed-wing aircraft may have decreased, the threat from TBMs, cruise missiles, and other unmanned aerial vehicles continues to grow. TBMs have many employment options. They offer various warhead choices, operate over extended ranges, and are relatively inexpensive.

(2) Theater missile defense (TMD) is a joint mission. It is accomplished by establishing an effective, interoperable battle management/command, control, communications, computers, and intelligence (BM/C^{4}I) system that permits the joint force commander to integrate and enhance the joint force's capabilities to—

- Destroy incoming theater missiles in–flight (active defense).
- Reduce the vulnerability of friendly force and critical assets from the effects of theater missile attacks (passive defense).
- Destroy hostile theater missile capability by offensive actions against missile launchers; command, control, communications, and intelligence (C^{3}I) and logistics facilities; and other theater missile infrastructure (attack operations).

(3) Army aviation plays a key role in TMD by executing deep operations to attack all elements of the hostile theater missile system.

(4) This mission could be conducted as a deliberate attack against known systems or as a search and attack operation when exact locations are not confirmed.

(5) Army aviation faces several challenges in the future execution of TMD to include airspace management, obtaining and processing real–time target information, and range of aircraft, while balancing payoff with mission risk. Although the challenges are many, Army aviation—with the AH–64 Longbow Apache-brings significant range, lethality, connectivity, and survivability to the TMD mission. Army attack helicopters offer unique and complementary potential to the ground commander when properly planned for and employed in TMD operations.

(6) Enhanced situational awareness through digitization also will provide important in–flight, divert capability to high–priority targets.

NOTE: As of the writing of this manual, tactics, techniques, and procedures for TMD are being developed, defined, and refined at all levels.

f. ***Special Operations.***

(1) Special operations aviation (SOA) units are trained, equipped, and manned to support both special and conventional operating forces. Special operations cover a series of unique primary, collateral, and emerging missions that directly support a theater combatant commander.

(2) Army SOA assets are dedicated to conducting special operations missions across the full range of military operations. They provide a mix of short-, medium–, and long–range lift, and limited light–attack capabilities. They support all principal, collateral, and emerging mission areas; they can conduct autonomous special reconnaissance and direct action missions.

(3) FM 1–108 contains detailed information on SOA.

g. ***Support by Fire.***

Support by fire (SBF) is a mission given to attack helicopters, directing them to establish a base of fire or an overwatch position. It can be used to engage a target while ground or air maneuver assets move to or bypass the same target area. It may range from suppression to destruction of the target; however, the primary mission is to fix the target so another force may maneuver. SBF positions are less restrictive than battle positions.

2-4. AVIATION COMBAT SUPPORT MISSIONS

Aviation combat support (CS) is the operational support and sustainment provided to forces in combat by aviation units.

a. ***Command, Control, and Communications.***

(1) Maintaining command, control, and communications (C^3) is critical to any operation. Aviation units provide communication enhancement through airborne transmission or relay equipment. Aviation assets, such as the A^2C^2S, permit commanders to quickly see their AO and command, control, and communicate on the move.

(2) Aviation assets may conduct liaison between separate units, transmit intelligence, and verify unit situations and locations. Other intelligence functions include target acquisition, reconnaissance, and employment of intelligence-gathering systems. The speed, flexibility, and communication assets inherent to aviation units contribute to the synchronization and deconfliction of Army combat forces.

b. ***Air Movement.***

Air movement operations are conducted to reposition units, personnel, supplies, equipment, and other critical combat elements in support of current and/or future operations. These operations include both airdrops and air landings. As these operations are usually aviation-pure missions, the aviation unit commander is usually the most qualified to produce the greatest efficiency of movement.

c. ***Electronic Warfare.***

(1) Electronic warfare (EW) is an essential component of C^2 warfare (C^2W). As part of C^2W, EW is used in conjunction with multidisciplined counterintelligence to protect friendly C^2 while attacking the enemy's C^2 structure. Effective use of EW—as a decisive element of combat power—requires coordination and integration of EW operations with the commander's scheme of maneuver and fire support plan. The integrated use of EW throughout the battlefield supports the synergy needed to locate, identify, damage, and destroy enemy forces and their structure.

(2) SEMA use the electromagnetic spectrum to locate, and target, enemy units and facilities; intercept enemy communications; disrupt enemy C^4I; and target acquisition capabilities. SEMA are organic to corps and divisions. They receive their mission taskings from the G2, not the aviation commander. Unmanned aerial vehicles (UAVs) may be assigned to aviation units but may also get their mission taskings from an external source. For both operational and safety reasons, both manned and unmanned aerial SEMA platforms must operate within the A^2C^2 system.

d. ***Combat Search and Rescue.***

(1) Aviation units must be prepared to conduct combat search and rescue (CSAR) in support of their own operations and to provide support at both the intra– and inter–service levels. CSAR planning should begin before forces deploy or immediately after arrival in the AO. Aviation units must develop a complete CSAR posture using a planning process that is fully complementary to ongoing operational planning. CSAR plans must be designed with the flexibility to employ all joint CSAR–capable resources in the most efficient and effective manner.

(2) For detailed planning of CSAR operations, refer to FM 1-111 and FM 90-18.

e. ***Air Traffic Services.***

(1) Air traffic services (ATS) encompass two areas: Army airspace command and control (A^2C^2) and air traffic control (ATC). ATS units provide a range of support that spans the entire theater during deep, close, and rear operations. Also, ATS operations span the wide range of military operations servicing Army, service component, interagency, multinational, and host nation airspace users.

(2) Specifically, ATS personnel support the A^2C^2 system, a subordinate element of the Army C^2 system. ATS liaison personnel, along with other staff representatives, are found at the division, corps, and theater A^2C^2 elements, as well as at other airspace-related elements within the theater air–ground system. They provide technical expertise in the operation of the A^2C^2 system to coordinate, integrate, and regulate use of a defined area of airspace by all users of that airspace. In addition, they integrate the division and corps airspace information centers—through which air operations data concerning friendly, unknown, and hostile aircraft are exchanged with subordinate units and the tactical operations centers (TOCs).

(3) ATC are those operations that provide advisory, procedural, and positive control at terminal locations and through en route coordination centers. These operations are both tactical and fixed base in nature, found from brigade landing/pickup zones to theater logistical airfields with full instrumented services. ATC units can conduct airborne, air assault, or air landing operations onto the battlefield; and immediately establish ATS throughout a theater. In many theaters Army ATS will be the first on the scene, and they will be controlling aviation forces from all services. ATS services include—

- Airspace deconfliction and airspace control measures.
- Navigational assistance.
- Flight following.
- Air threat warnings.
- Weather information.
- Artillery advisories.
- En route navigational structures.
- Airfield/landing zone. (LZ)/pickup zone (PZ) terminal control.
- Precision and nonprecision instrument approaches.

(4) For further information on these services consult FM 100-103.

f. ***Aerial Mine Warfare.***

Aerial-delivered mines can support tactical operations by emplacing tactical minefield; reinforcing existing obstacles; closing lanes, gaps, and defiles; protecting flanks; and denying the enemy AD sites. Aerial-delivered minefield also can be employed for flank protection of advancing forces and for operating in concert with air/ground cavalry units performing screen and guard missions.

2-5. AVIATION COMBAT SERVICE SUPPORT MISSIONS

Aviation combat service support (CSS) is the assistance provided by aviation forces to sustain combat forces. One aviation brigade can restore a mechanized battalion task force worth of combat power to a division each day through the expeditious movement of critical repair parts. Army aviation provides air movement of personnel, equipment, and supplies; and performs aeromedical evacuation and aviation maintenance.

a. ***Aerial Sustainment.***

Aerial sustainment is the movement of equipment, material, supplies, and personnel by utility, cargo, and fixed-wing assets for operations other than air assault and combat support. These air movements are considered CSS missions because the aviation forces are not task organized with combined arms forces, nor do they move CS forces or assets whose primary mission is to engage and destroy enemy forces. Missions include intratheater airlift; administrative relocation of troops and nonmilitary personnel; and administrative relocation of equipment, material, and supplies.

b. ***Casualty Evacuation.***

(1) Casualty evacuation (CASEVAC) is a part of combat health support. CASEVAC includes battlefield pickup of casualties; evacuation of casualties to initial treatment facilities; and subsequent movement of casualties to treatment facilities within the combat zone. CASEVAC is an aviation mission directly supporting a ground unit with casualty evacuation aircraft from forward locations to the brigade support area (BSA) or other designated collection/treatment facility. Aeromedical assets also will move medical personnel and supplies. Medical evacuation (MEDEVAC) is the process of moving patients while providing them en–route care. Most aviation units are not equipped or staffed to perform MEDEVAC. It is also requested through medical channels. CASEVAC can be performed by any Army aviation utility aircraft when tasked by the maneuver commander. These requests would go through aviation channels.

(2) FM 8–10–6 provides further details on MEDEVAC employment.

Section III. AVIATION OPERATIONS

2-6. OFFENSIVE OPERATIONS

Corps and division aviation assets will contribute during the preparation for offensive operations by assisting in finding, fixing, and engaging the enemy. When early engagement of enemy forces is desired in a meeting engagement, aviation forces may be employed to develop a situation until adequate ground forces can be moved into position to join in a hasty attack.

a. *Movement to Contact.*

(1) A movement to contact gains initial ground contact with the enemy or regains lost contact. Cavalry units, attack units, and target acquisition and reconnaissance units perform the movement to contact like a zone reconnaissance. Unlike a zone reconnaissance, the effort focuses on finding the enemy force; developing the situation early; and preventing the premature deployment of the main body following the cavalry. Terrain reconnaissance is conducted as necessary to support the intent of locating the enemy.

(2) As a result, movement to contact proceeds much faster than a zone reconnaissance. The division cavalry squadron can perform this mission when serving as part of a covering force or advance guard during a division movement to contact. A movement to contact is characterized by rapid, aggressive action. The commander rapidly develops the situation and may be permitted, particularly in division cavalry, to bypass enemy forces to maintain momentum. During a movement to contact, aviation assets may perform a number of tasks to include—

- Reconnoiter and determine the trafficability of all high–speed routes within the zone.
- Inspect and classify all bridges, culverts, overpasses, and underpasses along high–speed routes.
- Identify all bypasses and fords that cannot support rapid, heavy movement.

Find and report all enemy forces within the zone and determine their size, composition, and activity.

(3) The cavalry squadron gains contact using the smallest element possible. This is normally scouts or aeroscouts performing reconnaissance for their troop. Actions on contact occur rapidly at platoon and troop level to prevent unnecessarily deploying other squadron assets.

(4) Division cavalry facilitates speed by using air cavalry to reconnoiter forward of the ground troops or to screen along exposed flanks. The reserve allows flexibility on contact and rapid resumption of movement by the troops.

b. ***Attack.***

(1) During attack operations, aviation forces are employed in the close fight; they can be employed deep against second echelon forces, enemy artillery, helicopter forces, and enemy reaction forces, which could disrupt the momentum of the attack. Destruction of enemy C^2 nodes can also be critical to the success of the attack.

(2) Operations beyond the depth of the close fight—especially when conducted in synchronization with other combined arms, and joint service contributions—can break the cohesion of enemy defenses and lead to exploitation and pursuit. These operations are least effective against dug–in targets.

c. ***Exploitation.***

During exploitation operations, massed aviation assets under the aviation brigade may be used to maintain pressure on the disintegrating enemy forces. They also may be used to strike enemy forces attempting to reform or to provide reconnaissance in front of friendly advancing ground exploitation forces. Aerial reconnaissance gives the commander the capability to fight for information in the third dimension; then, operating in conjunction with ground forces, it can optimize the speed of advance.

d. ***Pursuit.***

(1) When an exploitation or pursuit scenario develops, the inherent speed and mobility of aviation forces are ideally suited to maintain enemy contact, develop the situation, and deliver aerial fires upon positions of enemy resistance. Since pursuit is a difficult phase of an operation to predict, forces may not be positioned to properly exploit the situation.

(2) Aviation forces may be moved quickly to find, fix, and attack fleeing enemy units; locate the enemy strike forces; and guide US ground forces into attack positions or around enemy exposed flanks. The maneuverability and firepower of Army aviation make it the optimum force to conduct both exploitation and pursuit operations.

e. ***Search and Attack.***

(1) Search and attack operations (a form of movement to contact) are generally conducted by smaller, lighter maneuver forces in densely forested areas to destroy enemy forces; deny area to the enemy; and collect information. They may also conduct search and attack operations—

- Against a dispersed enemy on close terrain unsuitable for armored forces.
- In rear areas against enemy special operations forces (SOF) or infiltrators.
- As area security missions to clear assigned zones.

Search and attack operations can prevent the enemy from planning, assembling, and executing operations on his own initiative.

(2) Most search and attack operations begin without detailed prior information about the enemy. The commander must produce much of his own intelligence as the operation unfolds. These operations are conducted at company, battalion, and brigade levels with division support. Historically, units conduct search and attack operations—

- In an environment of friendly air and fire superiority.
- Against squad–to–company size forces equipped with small arms and mortars, but normally without artillery support.
- Against both regular and guerrilla forces whose locations are unknown.
- In an environment where the enemy has the advantage of knowing both the terrain and the local populace.

(3) There is a significant risk associated with this mission. If the aviation unit is surprised by a well–prepared, dug–in force, its effectiveness drops drastically; the probability of aircraft losses increases significantly. FM 1–112 describes search and attack in greater detail.

2-7. DEFENSIVE OPERATIONS

In defensive operations, the speed and mobility of aviation are used to maximize concentration and flexibility. During preparation for defensive operations, Aviation may support the covering force with aerial reconnaissance and fires. During the defense, aviation can be used to attack deep against high–payoff targets, enemy concentrations, and moving columns; and to disrupt enemy centers of gravity. Attack helicopter battalions can be employed in depth to attack follow–on echelons before they can move forward to the close battle. Aviation forces can be employed to conduct screening operations; in conjunction with ground forces, they conduct guard operations on an open flank.

a. ***Mobile Defense.***

(1) The mobile defense is a defense that actively orients on the destruction of the enemy force. Generally, the force commander will resort to a mobile defense under the following conditions:

- Friendly forces are insufficient to adequately defend the AO.
- The commander possesses sufficient mobile forces to create a striking force.
- Orientation of the defense is for the destruction of the enemy force versus the retention of terrain.

(2) The mobile defense employs a combination of fire and maneuver, offense, defense, and delay to defeat the enemy attack and destroy the enemy force. The main effort in the mobile defense will be the striking force (Figure 2-3). Other considerations in a mobile defense might include—

- The planning of forward displacement of fire support assets when the striking force attacks.
- The ability of the defending force to provide fire support to the striking force to mass fires.
- The fact that the targets of the striking force may be beyond conventional artillery range.

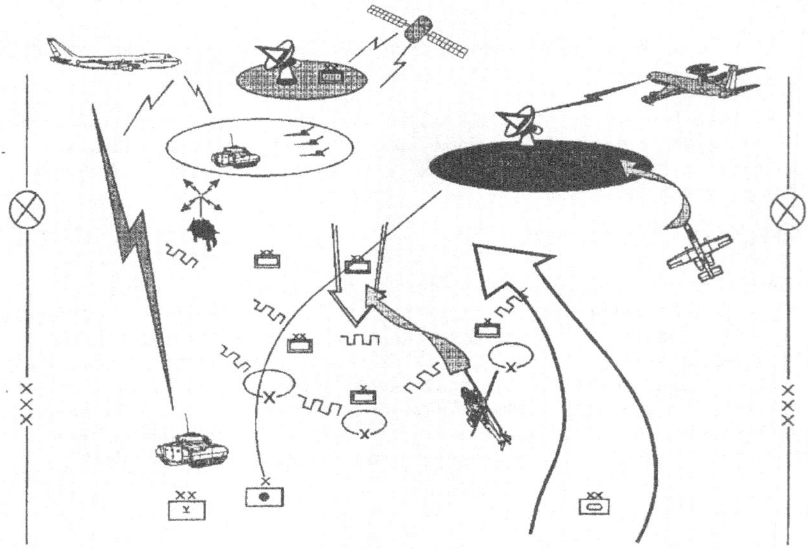

Figure 2-3. Mobile defense. As the enemy loses momentum, striking forces are available to plug gaps or counterattack to gain the initiative and begin transition to offensive operations. As lead enemy units are met by friendly units in well-established defensive positions, attack helicopters can engage follow-on or second echelon forces to slow or stop them from reaching friendly ground forces.

(3) The striking force is key to the commander's scheme of maneuver; thus, the mobile defense may fail without its commitment. It is not a reserve since it is deployed on a specific mission; it is not available for commitment elsewhere. The mobile defense normally will have a reserve independent from the striking force. Attack helicopter battalions can be used to blunt the enemy's attack; thereby, they assist in the setup for the striking force.

(4) During the striking force attack, aviation forces can support-with direct and indirect fires-the attacking maneuver force. Black Hawk and Chinook helicopter units can assist in moving artillery and infantry to support the striking force attack. Together, combat aviation and ground maneuver forces provide a much more effective strike force that can bring simultaneous fires to bear upon the enemy from unexpected directions.

b. ***Area Defense.***

(1) Area defense (Figure 2-4) is a defense that focuses on denying the enemy access to designated terrain or facilities for a specific time, rather than on the outright destruction of the enemy. The area defense is normally organized around static defensive positions in depth, seeking to destroy the enemy forces with interlocking fires.

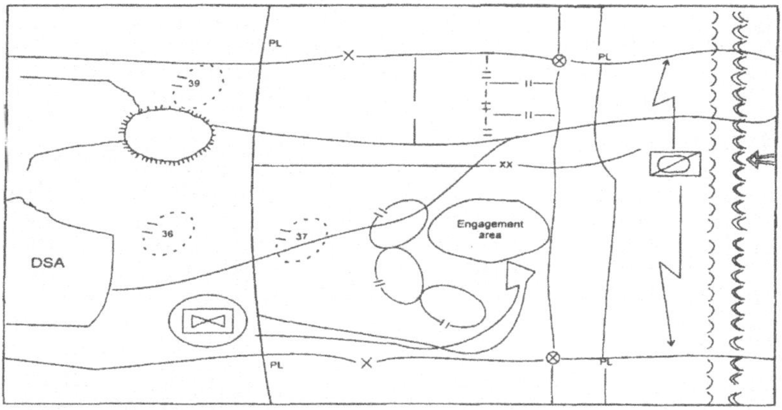

Figure 2-4. Area defense

(2) Division commanders normally position their forces in sectors and/or battalion battle positions on suitable terrain with a specific orientation of fires. In area defense operations, the ground commander can employ aviation maneuver forces to help contain

tactical emergencies—by disengaging them from one area and quickly concentrating them in another. Also, the aviation brigade's mobility and agility permit division and corps commanders to leverage risk by possibly eliminating the necessity of holding as large a ground maneuver force in reserve.

2-8. RETROGRADE OPERATIONS

a. In the conduct of the delay, aviation forces can assist the ground commander—by rapid concentration and employment of fires—to allow for disengagement and repositioning of friendly forces. Aviation forces can be employed to conduct surprise attacks to confuse advancing enemy formations. Air delivered mines can be used to supplement obstacles emplaced by engineers to impede or canalize enemy movements throughout the battle space. Air assault forces may be used to move rapidly between delaying positions.

b. The withdrawal, as in the delay, uses air cavalry and attack helicopters, in an offensive posture, to attrit enemy maneuver and fire support units; and to provide security for withdrawing friendly forces. During retirement, aviation forces can perform security operations to protect the movement of ground forces.

c. Retirement operations are conducted primarily at night; therefore, aviation's ability to maneuver, find, fix, and destroy the enemy, during the hours of darkness, is an advantage to the ground commander. Air cavalry units can assist in the security of routes of withdrawal. The retirement may occur over extended distances, and the security mission may be given to the corps or division aviation brigade commander. If so, appropriate ground units should be placed under his OPCON.

2-9. PLANNING

Although the planning focus for the corps differs from that of the division, the planning guidelines at both echelons are similar for aviation forces. Whether a corps commander is deciding on how to shape tomorrow's battlefield, or a division commander is planning tonight's counterattack, the planning principles for aviation brigades remain constant—brigades plan and battalions execute.

a. ***Deep Operations.***

(1) The aviation brigade provides attack battalions to destroy the high–payoff targets in the form of maneuver objectives selected by the force commander. Air assault units conduct deep operations to place infantry at critical areas of the battlefield in support of the scheme of maneuver. Aviation also inserts and extracts special operations forces and long–range reconnaissance teams. Deep operations (Figure 2-5a) require intensive detailed joint planning, coupled with extensive intelligence preparation of the battlefield (IPB).

(2) Some of the coordination functions that must reconducted for a successful mission are—

- Tracking the enemy through a series of named areas of interest (NAIs) and target areas of interest (TAIs).
- SEMA and Tactical Exploitation of National Capabilities (TENCAP) sensor support and product dissemination coordinated for use in real or near real time (NRT).
- Development of a decision support template (DST).
- Joint suppression of enemy air defense (JSEAD).
- Indirect fires.
- Friendly air defense artillery (ADA) status.
- Airspace deconfliction coordinated in the A^2C^2 cell.
- Synchronization with the ground scheme of maneuver.

(3) In a high–threat environment, aviation deep operations must be fully supported by elements of all the battlefield operating systems to ensure success. Long distances traversed over hostile territory will demand heavy emphasis on JSEAD.

(4) Use of cannon artillery, multiple launch rocket systems (MLRS), and Army tactical missile systems (ATACMS)—to suppress and destroy enemy forces along the route or in the target area—must be carefully synchronized. Army and Air Force EW capabilities should be planned as part of a deep operations package. Joint deception operations may be employed. Logistical aspects of the operation must receive heavy emphasis. Contingency planning facilitated by predictive intelligence will allow force packages to be prepared in accordance with the commander's intent. This should allow quick reaction to an execution "frag–order" (FRAGO) that can set the operation in motion in minutes, rather than hours. Application of the decide, detect, and deliver methodology will enable the aviation brigade to be responsive even when the demands of distance and enemy reaction make the planning complex.

b. ***Close Operations.***

(1) In close operations (Figure 2-5b), aviation is a great combat multiplier. Massing attack helicopters in the main effort greatly enhances the scheme of maneuver. Aviation monitors the division commander's battle and delivers near real–time intelligence. Aviation can rapidly shift focus and concentrate forces for critical engagements. Aviation units conduct maneuver—using the same standard maneuver graphics as ground forces—and fight from battle positions similar to armor and infantry.

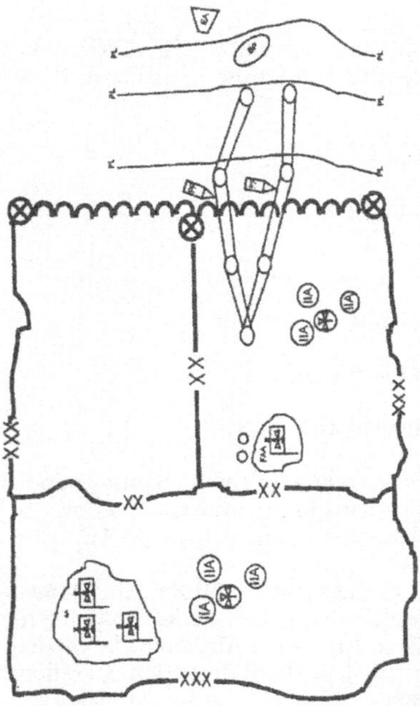

Figure 2-5a. Deep mission

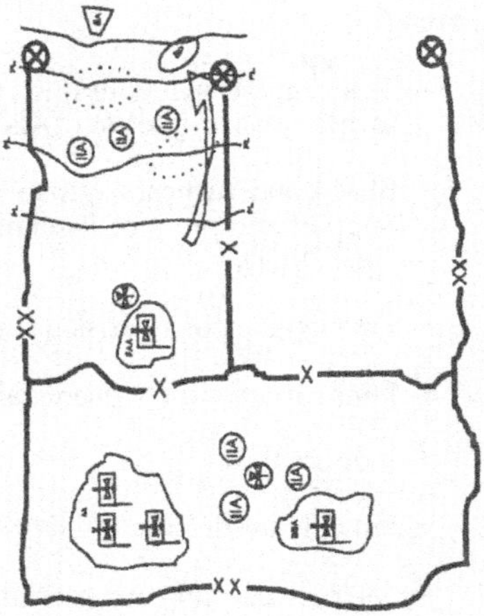

Figure 2-5b. Close mission

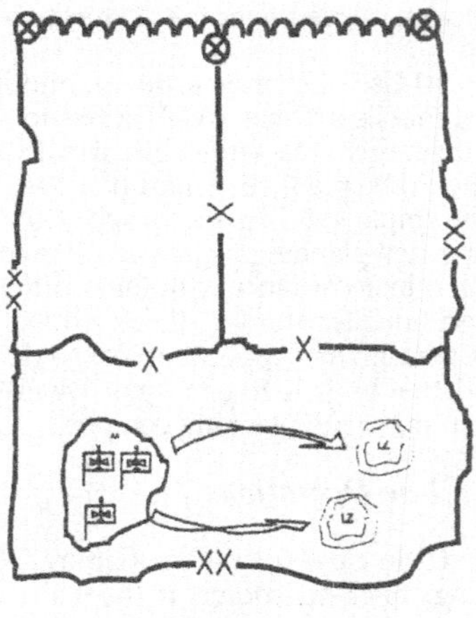

Figure 2-5c. Rear mission

(2) The principles of direct fire planning require the same terrain analysis principies for an AH–64 Apache company commander as that required of an M1A1 Abrams tank company commander. One difference, the Apache company commander can expect to operate throughout the entire corps/division AO; he must be prepared to execute operations anywhere within the corps commander's battlespace.

(3) Army aviation units may, on a mission basis, be placed under OPCON of a ground maneuver brigade. Conversely, infantry and armor units may be placed under OPCON of an aviation battalion or brigade. Because of the possibility of fratricide, it is imperative that such operations be carefully coordinated and that battlefield identification systems be used. Normal maneuver control measures are appropriate for controlling both aviation and ground forces.

(4) The coordination of airspace and fire support must be synchronized through A^2C^2 procedures. The effective use of combined arms maneuver in all areas of battlespace will help ensure the survivability of friendly forces. Aviation utility and C–47 Chinook aircraft provide essential CS and CSS within the main battle area. Army aviation assets shape the ground commander's fight by providing armed reconnaissance; critical C^2; rapid movement of combat power; EW operations; and delivery of aerial mines. CSS missions also support the fight by providing aeromedical evacuation; preplanned, and immediate, aerial resupply; air transport of exchange components; and pre–positioning of fuel and ammunition.

c. ***Rear Operations.***

(1) Aviation offers a full range of capabilities during rear area operations (Figure 2-5c). It can be a tactical combat force (TCF) or act as a reaction force against enemy threats. Aviation's providing aerial resupply, troop movements, aeromedical evacuation, and movement of equipment—such as artillery—across the battlefield is an example of the more common missions accomplished behind the forward line of own troops (FLOT).

(2) Detailed contingency planning is required for success in all rear area operations. Contingency planning and establishing C^2 relationships for the rear battle sets the conditions for commitments for aviation forces in the rear areas. The aviation commander selects forces for the conduct of rear area operations based upon METT–T and contingency orders from higher headquarters.

2-10. AIR COMBAT OPERATIONS

a. ***Air Combat in Deep Operations.*** During deep operations, aviation—

- Provides aerial security during air assault and attack helicopter operations.
- Must coordinate air combat planning with the ADA commander and other combined arms forces in the area.

- Must develop procedures for surviving an enemy air attack and then reconstituting aviation units.
- May require allocation of aviation forces exclusively for air combat roles.
- May require on–order missions for the majority of aviation forces. This allows the force commander to commit only the aviation assets necessary to counter the air threat while others continue primary maneuver or support missions.

b. ***Air Combat in Short Range Air Defense Operations (Close/Rear Battles)***

(1) During short range air defense (SHORAD) operations—

- Air combat operations provide for the protection of the combined arms team to include combat, CS, CSS units, and other high–value assets and locations.
- Air combat operations provide self-defense to aviation units.
- Assets must coordinate air combat planning with the ADA commander and other combined arms forces in the area.
- Attack and air cavalry assets may be tasked or diverted to perform air combat operations.
- Aviation may augment ground ADA forces in protecting vital assets in the rear area while they are repositioning or reconstituting.

(2) Although aviation is developing a credible air combat capability, aviation units should not normally be assigned areas of AD responsibility. Aviation can best use its maneuverability and firepower to augment ground AD at the point of enemy attack or to temporarily fill gaps in the maneuver force's ADA coverage.

c. ***Other Issues in Combat Operations.***

(1) Air combat operations may be critical in future wars. Air combat is the engagement or evasion of enemy aircraft. It is always a specified or implied mission when an air threat is predicted or present. Air combat operations—as a component of SHORAD in the close and rear battles—assist in protecting the combined arms force; providing self–defense; and augmenting tactical AD systems.

(2) Risk versus payoff does not warrant using Army aviation assets in a dedicated air combat role; therefore, Army aviation's response to an air threat is primarily defensive. Whenever possible, planned destruction of an air threat should be accomplished with ADA, field artillery, and/or Air Force assets. In addition, entry into an air engagement detracts from Army aviation's overall mission as a maneuver force.

(3) Aviation commanders must be prepared to support the force commander by conducting both offensive and defensive air combat operations. Air combat engagements will be short; victory will go to the side that can concentrate effective fires first.

(4) A major consideration for maneuver force commanders will be how to allocate adequate aviation forces to the ground scheme of maneuver while retaining force sufficiency to conduct on–order counterair operations. Aviation commanders tasked with multiple combat missions involving a significant air threat must be prepared to apportion a part of their force to conduct air combat operations. METT–T and IPB will determine the amount of combat power used for air combat missions and counterair requirements.

(5) Any armed helicopter can be called upon to execute air–to-air combat with any of its organic weapon systems. The weapon of choice will almost always be an air–to-air missile; however, if the mission profile does not include them, 30mm and 20mm, rockets, TOW, and Hellfire missiles can all be used in an air-to-air role.

(6) FM 1–112, Appendix F, includes detailed procedures and engagement criteria for each weapon system.

2-11. AVIATION LIAISON

a. Effective liaison between Army aviation units and supported elements is imperative. Aviation liaison officers (LNOs) will support maneuver, CS, and CSS operations. When under the OPCON of ground maneuver elements, aviation commanders should ensure that they are represented by well–trained, tactically proficient, LNOs especially during the planning process.

b. The role of the commander in this function cannot be overstated. Aviation commanders are the key linkage in establishing and perpetuating effective liaison; they should maintain a personal interface with the supported unit commander throughout operations.

c. LNOs, and S3/G3 air officers, must know aviation force structure; operational tactics; weapon systems capabilities, aviation maneuver employment; and sustainment requirements. The aviation LNO must be familiar with the capabilities of all Army aircraft available to the supported unit. These officers must continuously conceptualize how aviation can influence combat action and help other combat arms to achieve greater combat effectiveness.

2-12. STABILITY AND SUPPORT OPERATIONS (SASO)

a. ***Aviation Forces in SASO.***

(1) Aviation units have participated and can anticipate participating in every activity in SASO. Aviation's ability to rapidly deploy and operate effectively in austere environments makes it an invaluable asset in SASO. Aviation provides combat, CS, and CSS for SASO by—

- Reaching remote areas.
- Delivering food and medical supplies.
- Providing emergency communications.
- Providing aeromedical evacuation.
- Extracting disaster victims.
- Providing reconnaissance and security, combat projection, and the movement of personnel and equipment, administratively and tactically.

(2) The very presence of aviation makes it a highly visible deterrent force that can rapidly transition from peace to conflict.

b. ***SASO and Aviation Task Organization.***

(1) SASO can quickly transition from peacetime through conflict to war. Aviation units selected for SASO missions should carefully analyze the possibility of the mission deteriorating to conflict.

(2) A critical challenge facing aviation unit commanders is the task organization of their forces to accomplish the mission. As in the example of Somalia, a single aviation task force may well combine—under the operational control of one headquarters—the missions of the cavalry, attack, assault, CS, CSS, and humanitarian assistance. The aviation brigade commander, before deploying his force, must ensure that the deploying force is manned with a staff whose experience will cover the range of anticipated missions.

(3) Some major areas of consideration and planning include–

- Rules of engagement must be very clear to every commander and soldier; they should be specific enough to address the appropriate response to each known or suspected threat.
- Mission statement and commander's intent must be clear and understandable.
- Increased reliance on nonorganic personnel for assistance (nongovernment, civil affairs, counterintelligence).
- Mission versus METL: Identify training deficiencies before deployment.

- AD threat: tactics and techniques necessary to accomplish the mission.
- Military operations on urbanized terrain (MOUT) operations.
- Night vision operations in an urban environment.
- Combat search and rescue.
- Extraction of downed crews in urban terrain.
- Nondoctrinal service and support packages.
- Publications.
- Calibration, how/where.
- Maintenance sustainability.
- Compatibility with host nation/allies.
- Facilities.
- Cranes or other overhead lift.
- High operational tempo (OPTEMPO).
- Compatibility of aircraft to mission requirements.
- AVIM support.
- Combat identification.
- Aerial command post operations.
- Convoy security.
- Nondoctrinal communications requirements.
- Force protection.
- Psychological operations (PSYOP) operations.
- Transport of civilian personnel on military aircraft/rules and authority.
- Gunnery tactics, techniques, and procedures in an urban environment, limiting collateral damage, and eliminating fratricide.

- Flight following in areas without support infrastructure.

- Aircraft survivability equipment/survival vest, armor vests, weapons selection for crews, video cameras/video play back, rules of engagement, legal status of forces, limits of legal authority.

NOTE: The foregoing is not an inclusive list. The aviation commander, using all available information at hand, will have to anticipate requirements and organize his forces.

c. *SASO Overview.*

(1) SASO are designed to promote regional stability maintain, or achieve, democratic end–states; retain US influence and access abroad; provide humanitarian assistance to distressed areas; protect US interests; and assist US civil authorities. Such employment of Army forces may minimize the need for combat operations by defusing crises and nurturing peaceful resolution of issues.

(2) The Army usually conducts such operations as part of a joint team, and often in conjunction with other US and foreign government agencies. SASO are intrinsic to a combatant commander's peacetime theater strategy; an ambassador's country plan; or civil assistance, at home. The employment of aviation forces can be integrated by the combatant commander into the activities that support theater and country-specific plans to achieve regional and national objectives. Compel, reassure, and deter will be the foundation for SASO plans.

(3) SASO will not always have peaceful results. Several of the activities employing aviation forces will be conducted in the presence of hostile threat forces; they may result in combat, either by design or by the reaction of those threat forces.

(a) In general, the same principles and tenets that apply to aviation forces in combat operations will apply to aviation forces in these operations where the potential for combat exists. The main modification to the aviation principles and tenets is the need for restraint in SASO.

(b) In SASO, it is essential to apply appropriate military capability prudently. The actions of soldiers and aviation units are framed by the disciplined application of force in accordance with the specific rules of engagement. The use of excessive force could impede the attainment of both short–and long–term goals; therefore, restraints will often be placed on the weaponry, tactics, and levels of violence allowed in this environment. Also, because of the decentralized nature of operations often found in this environment, sergeants, warrant officers, and company grade officers are often placed in decision-making situations that could very well have strategic implications. Commanders should attempt to anticipate these situations and ensure the rules of engagement appropriately address them.

(4) The principle of security must be emphasized by aviation forces engaged in SASO. The presence of US forces in nations around the world may provoke a wide range of responses by factions, groups, or forces of unfriendly nations. Regardless of the mission, the commander must protect his force at all times. He should never be lulled into believing that the nonhostile intent of his mission does not put his force at risk. Inherent in this responsibility is the need to be capable of rapid transition from a peaceful to a combat posture, should the need arise.

(5) The activities in which aviation forces will be employed in SASO can be grouped in three main categories: *peacetime contingency operations; peacekeeping and humanitarian assistance;* and *military support to civilian authorities*. These categories are discussed below.

d. ***Security and Limited Conflict.*** The seven types of operations that have some potential to result in armed conflict are discussed below.

(1) ***Show of force.*** A show of force is a mission carried out to demonstrate US resolve in which US forces deploy to defuse a volatile situation that may be detrimental to US interests or national objectives. These operations can influence other governments or politico-military organizations to respect US interests and international law. They can take the form of combined training exercises; rehearsals; forward deployment of military forces; or introduction, and buildup, of military forces in a region. The mobility, flexibility, agility, and firepower of aviation forces make them ideal for employment in such operations. Emphasis in show of force operations for aviation will be on readiness to conduct combat and CS missions.

(2) ***Noncombatant evacuation operations.*** NEO relocate threatened civilian noncombatants from locations in a foreign country or host nation. These operations may involve US citizens whose lives are in danger; they may include selected host nation citizens or third country nationals. NEO may be conducted in the environments of conflict or war in a peaceful, orderly fashion or may require forcible means. Aviation forces are employed in the rapid air movement of noncombatants from endangered areas not safely served by fixed-wing aircraft. Scout and attack aircraft provide security for the air movement; they are prepared to engage hostile elements that may attempt to interfere with NEO.

(3) ***Counterdrug operations.***

(a) Military efforts support and complement—rather than replace—the counterdrug efforts of other US agencies, state agencies, and cooperating foreign governments. Aviation support can occur in any or all phases of a combined and synchronized effort to attack the flow of illegal drugs at the source, in transit, and during distribution. Army participation in counterdrug operations will normally be in support of law enforcement agencies. SEMA units play an active role in counterdrug operations.

(b) Support to host nations includes assistance to their forces to destroy drug production facilities; collaboration with host nation armed forces to prevent export of illegal drugs; and nation assistance to help develop economic alternatives to production, exportation, and distribution of drugs. Support to interdiction efforts centers on monitoring and detecting illegal drugs in transit as well as integrating C^3I systems. In interdicting drug production at the source, aviation units may be used to assist in locating production facilities; inserting reconnaissance, and special operations, teams; and supporting troop lift of indigenous forces engaged in counterdrug operations. Air cavalry scouts can be employed in the reconnaissance of suspected drug production areas, particularly at night, using forward-looking infrareds (FLIRs) and night observation devices. The contributions of aviation C^2 aircraft, assault helicopter units, and scout/attack aircraft can be effective in the conduct of such operations.

(c) Aviation units and soldiers may support domestic counterdrug operations in planning and providing training assistance. Equipment loans and transfers and other assistance may be requested and provided. This support may expand as national policy and legal restrictions evolve.

(4) ***Support for insurgences and counterinsurgencies.***

(a) At the direction of the National Command Authority (NCA), US military forces may assist either insurgent movements or host nation governments opposing an insurgency.

(b) US military resources will be used to provide support to a host nation's counterinsurgency operations in the context of foreign internal defense (FID) through logistical and training support. Military support to FID is provided through the unified CINC.

(c) Where US forces are supporting a host nation's counterinsurgency operation, most of the aviation missions can be employed effectively. Initially, aviation may assist host nation commanders with C^2 aircraft. Air assault units will be essential for tactical troop movements; scout or attack units may be required for reconnaissance and security. Aeromedical evacuation from remote or inaccessible locations may be required.

(d) In view of the uncertainty of counterinsurgency requirements, it may be that only a few aviation missions would be appropriate. In such a situation, it is possible that selected aviation battalions would be deployed and organized into provisional units, rather than deploying full aviation brigades. These deployments make it imperative that a modular concept of aviation logistics is in place to support nonstandard aviation organizations.

(e) The United States supports selected insurgences opposing oppressive regimes that work against its interests. Because support for insurgences is often covert, SOF are frequently involved. Their extensive unconventional warfare training and experience makes SOF aviation units well suited to provide this support. General purpose aviation forces may be employed when the situation requires their specialties or when the scope of operations is so vast that overt conventional forces are required.

(5) *Combatting terrorism.* The two major subcomponents to combatting terrorism are antiterrorism and counterterrorism. During peacetime, the Army combats terrorism primarily through antiterrorism–passive defensive measures taken to minimize vulnerability to terrorism. Antiterrorism is a form of force protection and is, therefore, the responsibility of aviation unit commanders at all levels. Counterterrorism is the full range of offensive measures taken to prevent, deter, and respond to terrorism. Army elements, such as SOF aviation units, assist in this interagency effort by applying specialized capabilities to preclude, preempt, and resolve terrorist incidents abroad. SOF aviation may conduct counterterrorism operations by inserting and extracting special operations teams and providing firepower to support their operations.

(6) *Peace enforcement.*

(a) Peace enforcement operations are military operations in support of diplomatic efforts to restore peace between hostile factions. These factions may not be consenting to intervention and may be engaged in combat activities. Peace enforcement implies the use of force or its threat to coerce hostile factions to cease and desist from violent actions. Units conducting peace enforcement must be prepared at all times to apply combat power to restore order, separate warring factions, and return the situation to one more conducive to civil order and discipline.

(b) Aviation units—which can be deployed into the area of operation with early entry ground forces-can have a significant deterrent effect on the indigenous combatants, particularly if these factions have armored forces. Air cavalry units or attack units may be employed to conduct reconnaissance and surveillance over wide areas and provide visual route reconnaissance. Chinook helicopter units may have an important role in moving military and civil peace enforcement personnel, or in delivering required supplies when warring factions interdict surface transportation or routes become impassable.

(c) Aviation forces employed in peace enforcement operations must operate in conjunction with ground maneuver forces that can interpose themselves between warring factions on the ground. Forces should expect ambiguous situations to be normal and must adhere to authorized rules of engagement. This is a difficult mission that requires restraint, patience, and a heightened awareness of force protection measures.

(7) *Attacks and raids.*

(a) The Army conducts attacks and raids to create situations that permit seizing and maintaining political and military initiative. Aviation is well suited to these combat operations because attacks and raids are normally conducted to achieve specific objectives other than gaining or holding terrain. Attacks by conventional air, ground, and aviation forces—acting independently or in conjunction with SOF—are used to damage or destroy high–value targets or to demonstrate US capability and resolve to achieve a favorable result.

(b) Raids are usually small–scale operations involving swift penetration of hostile territory to secure information; temporarily seize an objective; or destroy a target. Raids include a rapid, preplanned withdrawal after completion of the mission. Aviation forces conduct such attacks and raids using either attack helicopter or assault aviation assisted by air cavalry reconnaissance and security elements.

e. ***Peacekeeping and Humanitarian Assistance.***

(1) ***Peacekeeping operations.***

(a) Peacekeeping operations support diplomatic efforts to maintain peace in areas of potential conflict. They stabilize conflict between belligerent nations or factions; therefore, they require the consent of all parties involved in the dispute. Peacekeeping often involves ambiguous situations requiring the peacekeeping force to deal with extreme tension and violence without becoming a participant. As with peacemaking operations, aviation units and soldiers engaged in peacekeeping must apply restraint; have patience; and maintain a heightened security awareness, in executing these missions.

(b) Peacekeeping forces deter violent acts by their physical presence at violence–prone locations. They collect information on the situation by all means available. Scout aircraft and SEMA platforms are indispensable components of a joint peacekeeping force. C^2 and liaison aircraft will enable the leaders of the force to move to critical points rapidly and remain abreast of the situation as it develops.

(2) ***Humanitarian assistance and disaster relief.***

(a) Humanitarian assistance operations provide emergency relief to victims of natural or man–made disasters; they are initiated in response to domestic, foreign government, or international agency requests for immediate help and rehabilitation. Disaster relief operations include refugee assistance; food programs; medical treatment and care; restoration of law and order; damage and capabilities assessment; and damage control.

(b) Aviation can provide logistics support to move supplies to remote areas; extract or evacuate victims; assist in establishment of emergency communications; and provide aeromedical evacuation services, in support of medical operations. Aviation's ability to deploy rapidly—and its capability to operate effectively in austere environments—make it ideally suited for these missions.

(3) ***Nation assistance.***

(a) Nation assistance operations are conducted in support of a host nation's efforts to promote self–development. The goals of nation assistance will be specified in the ambassador's country plan and the CINC's theater strategy. The goals of nation assistance normally are accomplished through education and the transfer of essential skills to the host nation.

(b) Army aviation's participation in nation assistance will normally be limited to the use of individual soldiers and teams to train and educate; and the use of liaison aircraft to assist in overcoming terrain obstacles, and limited road nets, and as a means of communications.

(4) ***Security assistance.***

(a) Security assistance programs are the means by which the United States provides defense materiel, military training, and defense-related services-by grant, loan, credit, or cash sales—to further national policies and objectives. The main interface of the US Army with a host nation occurs through the Security Assistance Training Program. The program has two primary subcomponents—the International Military Education and Training Program (IMETP) and the Foreign Military Sales Program (FMSP).

(b) The IMETP is designed to enhance the proficiency, professional performance, and readiness of foreign armed forces. The Army conducts international education and training in the continental United States (CONUS), as well as in the host nation. The United States Army Aviation Center (USAAVNC) provides formal courses, orientation tours, and on–the-job training in support of this program.

(c) The FMSP allows designated governments to purchase military equipment, services, and training from the United States. The sale of aviation end items may require training in the operation and maintenance of those items. Mobile training teams, resident instruction at the USAAVNC and other US Army schools, and similar methods are used to conduct this training. The FMSP differs from the IMETP in that the recipient of FMSP pays for the equipment, services, and training received.

f. ***Military Support to Civilian Authorities (MSCA).***

(1) When appropriate governmental authority directs the armed forces to assist in domestic emergencies within CONUS, the Army has primary responsibility. Army aviation units support disaster relief; provide humanitarian assistance and ATS; and conduct similar operations, when directed. The ability of aviation units to rapidly deliver relief supplies and services to devastated or inaccessible areas rapidly is a critical advantage in the execution of such operations.

(2) Federal law authorizes the domestic use of military force to suppress domestic violence or insurrection. The Constitution and federal law, however, place restrictions on the use of military force in this manner.

(3) The Posse Comitatus Act requires specific Presidential or congressional approval and direction before Active Army or US Army Reserve forces may execute the law in place of duly appointed law enforcement means. The Army National Guard has similar, but less stringent, restrictions. In its capacity as a state militia, the National Guard may employ aviation units to assist state law enforcement officials largely at the discretion of the state government.

CHAPTER 3

ENVIRONMENTAL EFFECTS ON OPERATIONS

Weather information is critical to aviation planning. Aviation commanders and staffs must have current weather forecasts and observations throughout the entire area of operations (AO). Commanders must evaluate forecasted weather conditions for potential effects on plans, operations, and aviation assets. By exploiting adverse weather conditions, commanders and planners can take advantage of threat vulnerabilities and favorably influence their own courses of action.

3-1. EFFECTS OF WEATHER

Adverse weather can influence all Army aviation operations. Severe weather conditions may completely prohibit or drastically reduce aerial maneuver. However, some weather conditions benefit aviation operations. For example, fallen snow, extreme cold weather, or muddy terrain can limit or prohibit the movement of ground forces, while having little effect on aviation operations. Low cloud cover and ceilings may enhance aviation operations by yielding low–level cover and concealment. At the same time, these conditions restrict high–performance aircraft operations. Low–lying fog offers limited concealment from direct–view observation and weapons engagement.

3-2. GEOGRAPHICAL ENVIRONMENTS

Aviation conducts operations in all geographical environments. To effectively employ aviation forces, commanders must understand the unique characteristics of each environment and its effects on aviation operations as are described below.

a. *Mountains.*

Aviation forces are ideally suited for sustaining combat operations in mountainous terrain. In fact, the helicopter made its first real combat contributions in the mountainous terrain of Korea. Helicopters can overcome the difficulties associated with the movement and support of ground forces in mountains. Mountains provide excellent terrain-masking and radar and visual acquisition avoidance. Air movement is the principal mode for rapidly displacing forces, equipment, and supplies in mountainous areas where ground transportation is hindered from natural or man–made obstacles. The following factors affect operations in mountainous terrain:

- Terrain can limit maneuverability and engagement areas.
- In steep mountainous terrain, the look-down angle to a target can be so great that the target becomes almost impossible to engage.

- High altitudes restrict aircraft lift capabilities and armament loads.
- Weather conditions change rapidly. Snowstorms, strong winds, and turbulence occur in passes and over jagged terrain.
- Aircraft icing is common in high altitudes and may occur suddenly. Icing impedes lift in helicopters and can prevent them from flying. Most modern helicopters have de-ice/anti-ice capabilities for their rotor blades. However, these same icing conditions can still prevent attack helicopters from firing their weapons.

b. ***Jungles.***

(1) Jungle operations are characterized by dense vegetation, high temperature, high humidity, and heavy rain. Army aviation significantly enhances combat operations in jungle areas. Rugged terrain, dense vegetation, and lack of roads make Army aviation's combat, combat support (CS), and combat service support (CSS) critical in conducting jungle operations. Aviation effectively provides aerial fires, reconnaissance and security, air assaults, command and control (C^2), resupply, and aeromedical evacuation.

(2) The following factors affect aviation operations in jungle terrain:

- Range and effects of weapons systems are often limited by dense jungle vegetation.
- Artillery fire can be difficult to observe and adjust.
- Thick foliage and rugged terrain reduce the range of radio communications.
- Hot, humid tropical air decreases aircraft lift capabilities.
- Weather is subject to rapid and violent change.
- Problems with corrosion are intensified, thereby increasing demands on spare parts and maintenance.

c. ***Deserts.***

(1) Aviation forces can operate effectively in desert environments. Desert terrain often allows aircraft to engage targets at the maximum range of their weapons systems. However, aircraft are vulnerable to enemy long–range observation and detection because of inadequate cover and concealment.

(2) The following factors affect aviation operations in desert terrain:

- Reflected sunlight from aircraft canopies increases the possibility of enemy detection.
- High daytime temperatures decrease lift capabilities.
- Dry, sandy terrain degrades radio effectiveness.
- Windblown sand and dust cause increased maintenance and supply requirements as well as backscatter challenges for laser-designated systems.
- Featureless terrain complicates navigation.
- Telltale dust signatures and "brownout" upon landing.
- Rearm and refueling operations are complicated by blowing sand and dirt.
- Featureless desert terrain makes night vision goggle operations extremely difficult.

d. ***Arctic Areas.***

(1) Aviation combat, CS, and CSS operations should be considered normal operations in the arctic. Commanders must understand and appreciate how the polar environment affects aviation operations. In the arctic, operations maybe slowed by adverse weather or other environmental factors.

(2) Commanders operating in arctic or near-polar regions need to be aware of the following unique difficulties that aviation encounters in these regions:

- Severe arctic weather dictates major changes in operating procedures; considerations include operating in continuous darkness with low clouds, reduced visibility, whiteouts, and severe icing.
- Navigation is complicated by the scarcity of ground references.
- Arctic temperatures increase maintenance requirements.
- Aircraft capabilities may be reduced because of the additional equipment (such as skis) required to operate in arctic conditions.
- Aviation operations in snowy arctic environments may leave telltale snow signatures.
- Increased emphasis must be placed at all levels to protect the force from cold weather casualties.

e. *Urbanized Terrain.*

Urban operations present unique and complex challenges to aviation units. Urban operations can occur in any of the geographical environments. The following factors affect wiation operations in the urban environment:

- Restricted/limited landing zones (LZs) and pickup zones (PZs).
- Increased tower, antenna, and wire hazards.
- Foreign object damage to aircraft from flying debris.
- Operating in areas with high concentrations of civilians.
- Collateral damage to property.
- Night vision system operations in the vicinity of city lights.
- Degraded communications.
- High risk to aircraft from close–range, small arms fire, complicated by proximity of noncombatants.

3-3. NBC ENVIRONMENT

a. A major consideration for the force commander is the use of nuclear, biological, and chemical (NBC) weapons against friendly forces. These weapons through use, or threat of use, can cause large-scale shifts in tactical objectives, phases, and courses of action.

b. Aviation units can operate in the NBC environment. However, individual and unit performance is degraded when operations are conducted in mission-oriented protective posture (MOPP) 4. Routine tasks such as maintenance, aircraft/vehicle operation, locating and identifying targets, and night operations become infinitely more difficult when conducted in MOPP 4. The degradation caused by aviation units operating in MOPP can be significantly improved by conducting the actual operation in field training exercises while in MOPP gear. All aviation unit operations must be routinely practiced while in MOPP gear for this degradation to be overcome. It is particularly important for the pilot to be able to perform his or her combat mission while in MOPP gear.

c. The avoidance of contamination is essential for successful operations when faced with an NBC threat. Avoiding contamination allows aviation units to maintain the tactical momentum and preserves combat power by keeping soldiers out of increased NBC protective measures.

d. Force protection is an imperative in this environment. Commanders can ensure unit survivability by—

FM 1-100

- Preparing the unit for NBC operations.
- Establishing decontamination priorities/procedures.
- Dispersing forces.
- Ensuring units use NBC contamination avoidance procedures.
- Placing units outside enemy chemical weapons range.
- Using terrain for shielding against effects of NBC weapons.
- Establishing improved positions.
- Establishing NBC protective (MOPP) requirements and procedures.
- Camouflaging and using concealment properly.
- Integrating smoke operations into the scheme of manuever.
- Ensuring NBC detectors are properly employed.
- Ensuring adequate planning has taken place to defend against a biological attack.
- Considering the level of NBC training when planning NBC aviation operations. Units deficient in NBC training should receive additional training.
- Ensuring helicopters are provided to corps units requiring heliborne C^2 and surveillance assets.

e. Because of the unique requirements and challenges of nontactical NBC hazards by aviation elements during the support of military SASO, the ability to avoid contamination can and will be extremely limited. The force commander and the aviation elements must ensure that—

- Some method and location for decontamination of aviation assets are emplaced and operational.
- The spread of contamination in noncontaminated areas by overflights and/or decontamination operations is limited.
- All procedures to limit the effect on aircrews and passengers are emplaced (i.e., sand bags on floor to reduce radiation exposure).

3-4. ELECTRONIC WARFARE ENVIRONMENT

The worldwide proliferation of electronic warfare (EW) systems—systems that can automatically locate emitters through DF (direction finding (radio)), intercept, and jam our electronic signals—makes this an important environment for commanders to be aware of and plan for. EW may be present in any operational environment ilom SASO to war. Electronic attack is especially significant to the growing number of aircrafi that rely on electronic signals for flight control. For more information on aviation operations in an EW environment, refer to FM 34-25-7 (S).

CHAPTER 4

SUSTAINMENT OF AVIATION OPERATIONS

Sustainment of aviation operations is a command responsibility critical to the survival of the force and the conduct of combat operations. Sustainment is the responsibility not only of the aviation commander, but of commanders throughout the chain of command. Aviation commanders sustain their forces through four fimdamental elements: Force protection, logistical support, maintenance support, and personnel support. This chapter discusses these elements as well as risk management (a subset of force protection).

4-1. FORCE PROTECTION

a. The most critical element in the sustainment of aviation operations is *force protection.* It must be clearly understood by all commanders that aviation forces should be augmented with security forces to assist in protection of assembly areas and forward arming and refueling points (FARPs). Although the immediate consideration in force protection will be the security of the forces, it must also be understood that force protection applies to safety, maintenance, and the environment.

b. Aviation forces must be positioned at sites where they can be given adequate security support from ground commanders. This can be accomplished in several ways as follows:

- Revetment—bunker–base camp techniques such as used in Vietnam.
- Secure airfields away from hostile areas.
- At laager and battle position sites protected by distance from enemy forces because of the advance of our own ground combat elements.

c. The same considerations for risk analysis and accident prevention that would be applied at home station during peacetime operations also are critical to the protection of the force in combat operations. Maintenance procedures, crew rest, safety at all levels, increased operational tempo (OPTEMPO), increased demands on supply parts, increased maintenance workload, and environmental considerations must be constantly evaluated.

d. During combat operations, aviation forces derive their protection from the proper use of tactics, techniques, and procedures. Proper use of these elements will allow the aircraft to arrive safely on station to engage the enemy.

4-2. RISK MANAGEMENT

a. Another critical element of force protection is *risk management.* Risk management identifies actions that could help commanders eliminate, reduce, or minimize risk while maximizing force protection.

b. The risk management process identifies hazards that present the highest risk to the mission or operation. The process begins with mission analysis. When wargaming a course of action, planners identify hazards, assess the level of risk, and develop risk reduction options before they go to the next action. From this analysis, the staff considers the conditions most likely to cause mission failure and accidents.

c. For more information on command and staff responsibilities and the five-step risk assessment planning process, see Appendix F.

4-3. LOGISTICS PLANNING

a. The change in the present fighting posture to that of a United States–based *force projection* Army places greater demands on the aviation logistician. Extended supply lines, the requirement for continuous sustainment, and the need for intensive maintenance of ever–complicated systems demand a redundant and mobile logistical structure manned by expert soldiers trained in their particular skills.

b. The aviation commander must plan and monitor the logistical support to his force from predeployment through mission accomplishment and redeployment to home station. The aviation force commander can ensure the combat potential of his force only through effective use of the logistics principles.

c. Detailed logistics planning and coordination are essential to adequately support aviation operations. Aviation commanders must anticipate and forecast all logistical requirements based upon the anticipated OPTEMPO.

d. The basic rule is that the greater the operational demands—the greater the flying hours per airframe, amount of maintenance required, and amount of logistical support required. The aviation commander must consider—

- Sources and adequacy of support and distance of his force from his support base.
- Transportation availability allocated to support aviation resources—time, distance, and whether or not support will be continuous and uninterrupted.
- Communication links to supply source—wire, radio, or satellite communications (SATCOM). Is a computer link available, possible, or required?

- Are adequate ground transportation resources allocated to the aviation commander to transport the increased parts stockage required by increased OPTEMPO?

- Will roads be passable or will they be clogged with refugees significantly slowing ground resupply?

- Where will aviation fuel be obtained; how will it be transported; what is the distance from fuel site to operational area; what is the ability to test fuel for contaminates; what alternate fuels are available, etc.?

- Support facilities available in operational areas, i.e. overhead lift, machine shops.

- Special environmental considerations such as fuel, oils, and lubricants for arctic environments; protection kits for sand and dust particles.

e. The preceding items are not all encompassing; they highlight only the complexity of aviation logistics when deploying to hostile and austere environments.

4-4. MAINTENANCE SUPPORT

a. The aviation force commander must ensure that aviation maintenance support units repair and return aircraft to a serviceable condition as rapidly as possible. These tasks require aviation maintenance organizations to conduct 24–hour, continuous maintenance operations. Aviation force commanders and aviation maintenance commanders must work closely to plan and coordinate aviation equipment repair for return to service as far forward as possible.

b. Aviation support requirements are a function of the total time necessary to recover and perform essential repairs. Depending on the level of repairs required, aviation unit maintenance (AWM) or aviation intermediate maintenance (AVIM) support teams may be sent forward to repair battle-damaged aircraft. They will attempt to make those minimum repairs necessary for the aircraft to continue its mission or to enable it to be flown to a secure location where additional maintenance can be performed.

c. Downed or inoperable aircraft may need to be recovered by ground or air assets for repair operations. In either case, aviation maintenance and aircraft repairs are conducted as far forward as possible; self–recovery is preferred with aerial and ground recovery methods used as a last resort.

d. As with other areas of logistics, the commander must remember that the increased OPTEMPO will increase the demand on his maintenance assets. Maintenance crews will be required to work harder and longer, and may, as in the case of Somalia, find themselves in a situation where the length and OPTEMPO of the mission have exceeded the ability of the maintenance units to return aircraft to a flyable status in a timely manner.

e. Under these conditions, time is the unchanging variable and only a decrease in OPTEMPO or an increase in maintenance personnel can solve the problem. If the aviation unit is deploying either to multiple operational areas or to a location a great distance from its higher level of maintenance support, then careful consideration must be given to the modularity and redundancy of specialized tools and test kits required to conduct aviation maintenance.

4-5. PERSONNEL SUPPORT

a. Personnel support is received from various division, corps, and theater–level units and elements. Primary staff coordination is provided by the theater and corps G1, the division G1/Adjutant General, and the brigade and battalion S1.

b. The aviation commander must ensure that—

(1) Combat critical personnel functions are conducted. These functions include replacement operations; strength management; personnel accounting and strength reporting; casualty management; personnel data base management; and postal operations.

(2) Other personnel actions are provided as the combat situation permits. These include promotions; evaluations; leaves and passes; identification documents; classifications; personnel assignments; and awards.

(3) Other personnel support is provided as required. This includes finance services to support local procurement and provide pay services; legal services; and chaplain activities.

4-6. ARMING AND REFUELING

a. The aviation brigade commander must ensure that the resupply of critical aviation fuel and ammunition is anticipated, planned, coordinated, and synchronized. The aviation brigade will establish FARP sites. Division and corps assets will throughput Class III and Class V to them.

b. Theater/corp utility and cargo helicopter assets also can provide emergency aerial resupply of fuel and ammunition. The aviation commander ensures rearmh-efuel locations are accessible to air and ground transportation assets. He must guard against enemy activities that may hamper rearming and refueling operations and also move periodically to ensure survivability.

4-7. GROUND VEHICLE OPERATIONS

Aviation commanders must realize that most of their support is not air–transportable. Therefore, ground wheeled vehicles are needed to support close and rear operations. These vehicles will transport much of aviation's fuel, ammunition, and maintenance support. Aviation units are high–priority targets for threat forces, whose targeting and engagement systems have become increasingly sophisticated. Aviation commanders must plan for security and protection of their assets so that operations can continue uninterrupted.

4-8. AVIATION RECONSTITUTION

a. Reconstitution is extraordinary action that commanders plan and implement to restore units to a desired level of combat effectiveness commensurate with mission requirements and available resources. Reconstitution is first and foremost an operational decision. Aviation commanders must recognize when their aviation forces are becoming combat ineffective. At that point, they must take specific actions to return forces to combat effectiveness.

b. The aviation force commander identifies the extent, distribution, and specific types of personnel and equipment losses. He also assesses residual combat capabilities. He then must coordinate with the theater, corps, or division commander and staff concerning needed replacements.

c. Two major elements of reconstitution are reorganization and regeneration. Reorganization is action to shift resources within a degraded unit to increase effectiveness. (Combining two or more attrited units to form a single mission-capable unit is one example.) Regeneration is the rebuilding of a unit through large-scale replacement of personnel, equipment, supplies, reestablishment of the chain of command, and training.

APPENDIX A

ORGANIZATIONAL STRUCTURES

A-1. ECHELONS ABOVE CORPS (EAC) AVIATION BRIGADES

a. Mission.

(1) EAC aviation brigades (Figure A-1) support EAC, corps, and division operations. Based on organizational structure, EAC brigades conduct combat, combat support (CS), combat service support (CSS) according to theater operational requirements. EAC aviation brigades conduct joint, combined, and combined arms maneuver operations, and support theater special operations forces (SOF).

(2) EAC aviation units primarily support subordinate tactical units in the corps and divisions, although EAC units may be required to conduct theater rear area security. These units may also serve as a tactical reserve.

b. ***Structure.*** An EAC brigade is designed, tailored, and configured for the specific theater in which it operates. EAC brigades are organized by the Army component commander based on the mission guidance from the theater commander–in-chief (CINC). The brigade may be organized with any combination of attack, reconnaissance, assault, CH–47 Chinook, fixed-wing, and maintenance units; however, the normal structure would include a headquarters and headquarters company (HHC); one UH–60 Black Hawk–equipped command aviation battalion (CAB); one fixed-wing theater aviation battalion; one CH-47 Chinook–equipped theater helicopter battalion; and one air traffic services (ATS) battalion. The theater Army area command (TMCOM) will provide an aviation intermediate maintenance (AVIM) company to each EAC brigade.

A-2. CORPS AVIATION BRIGADE

a. ***Mission.***

(1) The corps aviation brigade (Figure A-2) primarily conducts attack, reconnaissance, security, air assault, command and control (C^2), and air movement operations throughout the corps area of operations (AO). The corps aviation brigade plans, coordinates, and executes aviation operations in support of the corps scheme of maneuver; it can be expected to operate anywhere in the corps area. Attack helicopter units destroy enemy forces by fire and maneuver. Assault and CH–47 Chinook helicopter units transport combat personnel, supplies, and equipment for corps operations. Helicopters are provided to corps units requiring heliborne C^2 assets. ATS are provided for Army airspace command and control (A^2C^2) integration, airspace information, and terminal and forward area support services.

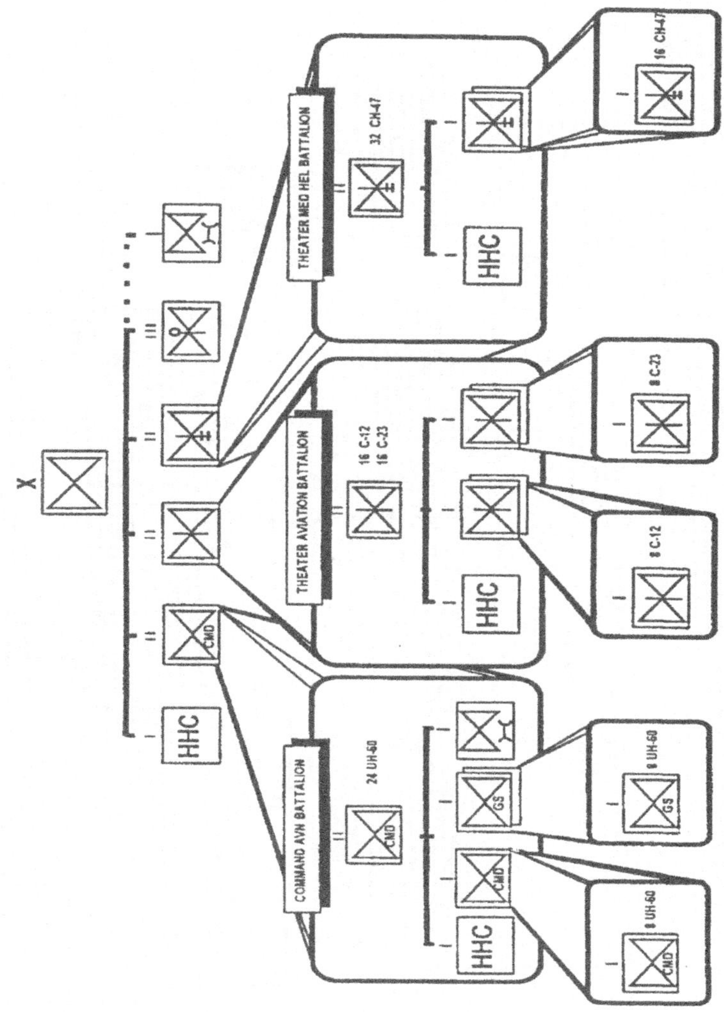

Figure A-1. Theater aviation brigade

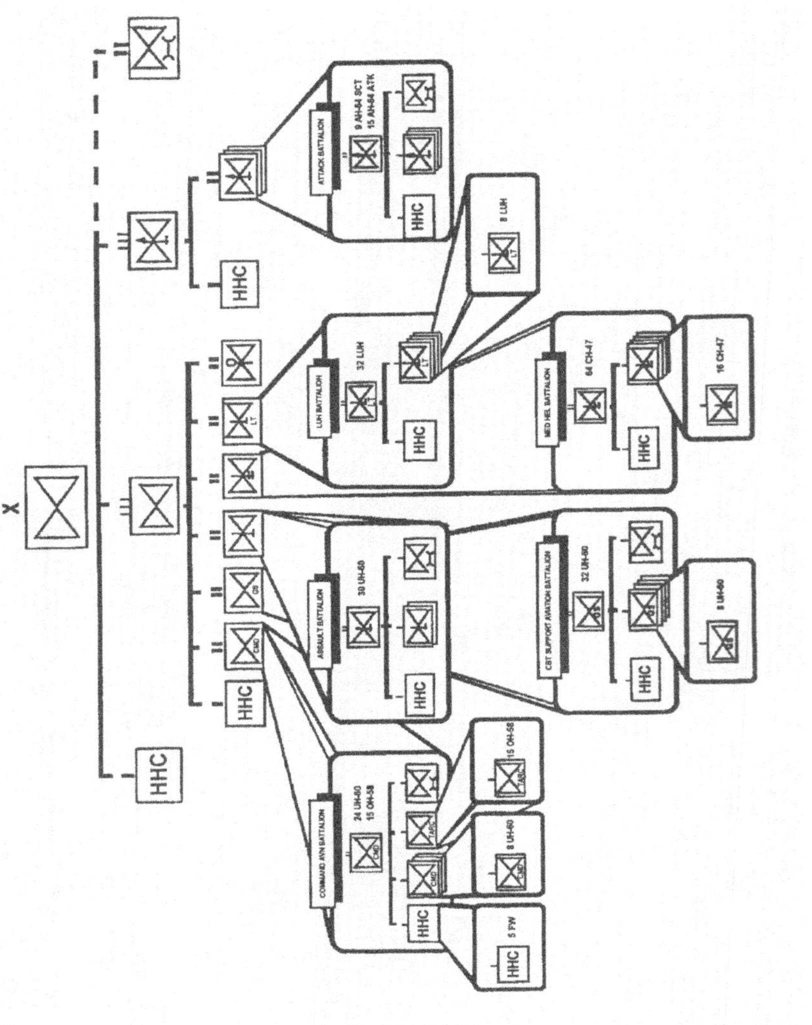

Figure A-2. Corps aviation brigade

(2) The corps atiation brigade mayoperate directly forthecorps comander or be placed under operational control (OPCON) of a subordinate division. The corps commander can task organize other corps assets under the command of the corps aviation brigade or task organize the corps aviation brigade to support an armored cavalry regiment (ACR). In corps without ACRs, the aviation brigade may be tasked to be the covering force headquarters.

b. ***Structure.*** Aviation brigades are organic to all US Army corps. Among the corps, there will be some differences concerning composite unit (COMPO) designations, but the organizations are basically similar throughout the Army. The corps aviation brigade is composed of one HHC, one aviation group, and one attack regiment.

(1) The aviation group consists of an aviation group HHC; an assault helicopter battalion (AHB); a command aviation battalion (CAB); a combat support aviation battalion (CSAB); a CH-47 Chinook helicopter battalion; a light utility helicopter (LUH) battalion; and an assigned ATS battalion.

(2) The attack regiment consists of a regimental HHC and three AH-64 Apache attack helicopter battalions (ATKHBs). Corps attack regiments receive C^2 support from the corps CAB, while CS and CSS are provided from the corps CSAB.

(3) Corps support command (COSCOM) provides an AVIM battalion to support the corps aviation brigade.

c. ***Utility helicopter support.*** The aviation brigade commander is key to the integration of the corps' aviation resources. The CAB, CSAB, and LUH support the aviation brigade as well as the corps with utility aircraft. The CSAB and LUH battalions also will provide CSS to the brigade's CH-47 Chinook battalion, AWM units, and the attached AVIM battalion. Habitual relationships, standard procedures, and regular training will facilitate this integration. The brigade staff will routinely plan and coordinate CAB and CSAB support for the attack battalions. In addition, the brigade commander will ensure that programs are established that reenforce collective training proficiency between the CAB/CSAB/LUH and the units they are required to support.

A-3. DMSION AVIATION BRIGADES

a. ***Mission.***

(1) The division aviation brigade conducts all aviation combat, CS, and CSS missions (except ATS and fixed-wing operations) in support of the division scheme of maneuver. The primary mission of the division aviation brigade is to find, fix, and destroy enemy forces within the division area. The division aviation brigade can accomplish this mission as an aviation–pure or task–organized force.

(2) The division atiation brigade may be supported by higher echelon aviation assets. The division aviation brigade must be prepared to support these assets throughout the duration of any given operation.

b. ***Structures.*** Aviation brigades are organic to all Army divisions. The designs of the aviation brigades have been tailored to meet the specific needs of the parent division, whether it be heavy, light, airborne, or air assault.

(1) ***Heavy division aviation brigade.*** The primary mission of the heavy division aviation brigade (Figure A-3) is to engage and destroy threat armored and mechanized forces. The heavy division aviation brigade is composed of a brigade HHC; one divisional cavalry squadron; one general support aviation battalion (GSAB); and two ATKHBs.

(a) The heavy division cavalry squadron consists of an HHT, an AVUM troop, three armored cavalry troops, and two air cavalry troops (ACTs). The two ACTs will be equipped with either eight OH–58D Kiowa Warriors or eight AH-1 Cobras each. In either case, the ACTs will be aircraft pure.

(b) The heavy division GSAB consists of a battalion HHC; a command aviation company (CAC) with eight UH–60s; six OH–58A/C Kiowas; and three EH–60 Black Hawks; two support aviation companies (SACs) with eight UH–60s each; and an AVUM company.

(c) The heavy division requires two ATKHBs. The attack battalions are AH-64s or AH-1 pure. An attack battalion consists of an HHC; an AWM company; and three attack helicopter companies (ATKHCs) equipped with eight attack aircraft each.

(d) The division support command (DISCOM) supports the division aviation brigade with an aviation support battalion (ASB) AVIM unit. The ASB has no organic aircraft.

(e) ***Utility helicopter support.*** The aviation brigade commander is the primary integrator of aviation assets within the division. The GSAB will provide support to the division, aviation brigade, cavalry squadron, the attack battalions, and the ASB. The aviation brigade will allocate resources based on mission, enemy, terrain, troops, and time available (METT–T); the scheme of maneuver; availability of assets; and the priorities set by the brigade commander. In addition, the brigade commander will ensure that programs are established that reinforce collective training proficiency between the GSAB and the units they are required to support.

(2) ***Light division aviation brigade.*** The primary mission of the light division aviation brigade (Figure A-4) is to conduct combat operations against light threat forces. The light division aviation brigade is composed of a brigade HHC; one divisional cavalry squadron; one AHB; and one ATKHB.

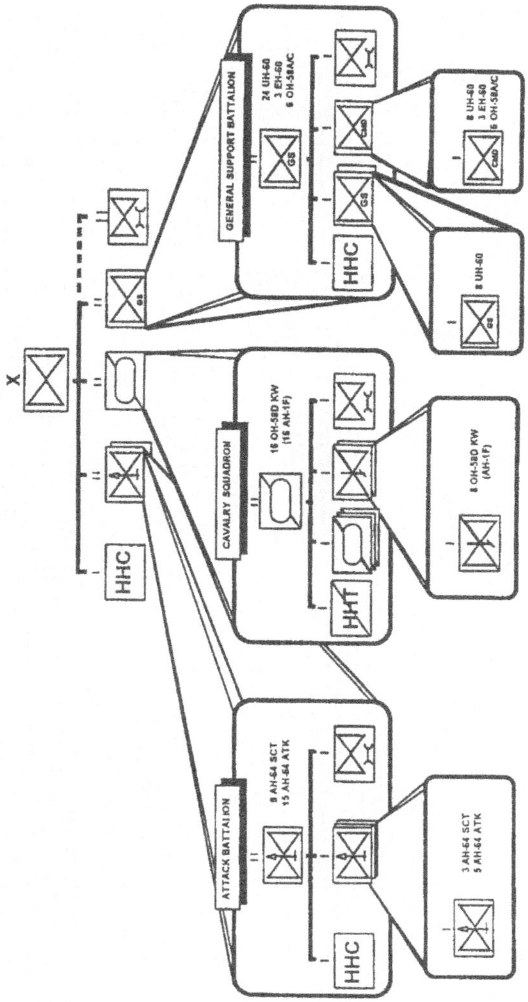

Figure A-3. Heavy division aviation brigade

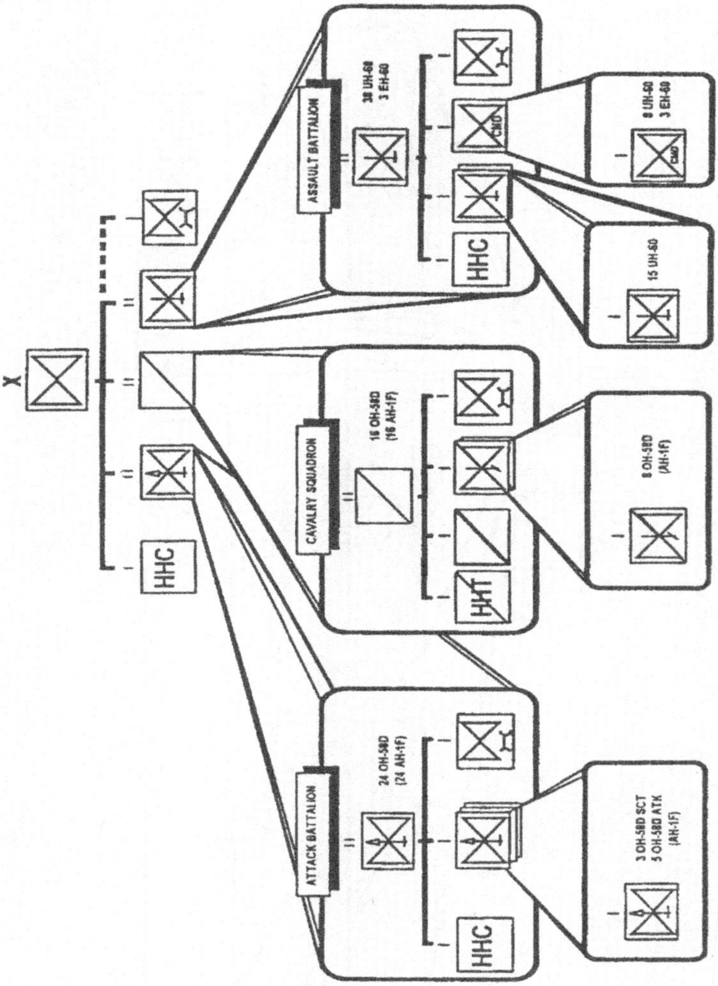

Figure A-4. Light division aviation brigade

(a) The light division cavalry squadron consists ofan HHT, an AVUM troop, one ground cavalry troop, and two ACTs. The two ACTs will be equipped with either eight OH-58Ds or eight AH-1s each. In either case, the ACTs will be aircraft pure.

(b) The light division AHB consists of an HHC, an AWM company, a CAC with 8 UH–60s and 3 EH–60s, and 2 assault companies with 15 UH–60s each.

(c) The light division ATKHB consists of an HHC, an AWM company, and three ATKHCs equipped with 8 OH–58Ds or 8 AH-1s each.

(d) The DISCOM supports the division aviation brigade with an AVIM company. The AVIM company has no organic aircraft.

(e) Utility helicopter support. The aviation brigade commander is the primary integrator of aviation within the division. The assault battalion supports the aviation brigade as well as the division with utility aircraft. The attack battalion, cavalry squadron, and the attached AVIM company will be supported by the assault battalion for C^2, CS, CSS, and CSAR. The brigade commander will ensure that programs are established to reinforce collective training proficiency between the AHB and the units they are required to support.

(3) Airborne division aviation brigade. The primary mission of the airborne division aviation brigade (figure A–5) is to rapidly deploy anywhere in the world and conduct combat operations. These combat operations will usually include link–up with follow-on forces at a later time. The airborne division aviation brigade is composed of a brigade HHC, one divisional cavalry squadron, one AHB, and one ATKHB.

(a) The airborne division cavalry squadron consists of an HHT, an AWM troop, one ground reconnaissance troop, and three air cavalry troops (ACT) with eight 0H–58Ds each.

(b) The airborne division AHB consists of an HHC, an AWM company, a CAC with 8 UH–60s and 3 EH–60s, and two assault companies with 15 UH–60s each.

(c) The airborne division ATKHB consists of an HHC, an AWM company, and three ATKHCs equipped with eight OH–58Ds each.

(d) The DISCOM supports the aviation brigade with an AVIM company. The AVIM company has no organic aircraft.

(e) Utility helicopter support. The aviation brigade commander is the primary integrator of aviation within the division. The assault battalion supports the division as well as the aviation brigade. Having no utility aircraft in the attack battalion requires brigade to support the moving of Class I, III, and V and resources. The ATKHB, the cavalry squadron, and the attached AVIM company will be supported by the assault battalion for C^2, CS, CSS, and CSAR. In addition, the brigade commander will ensure

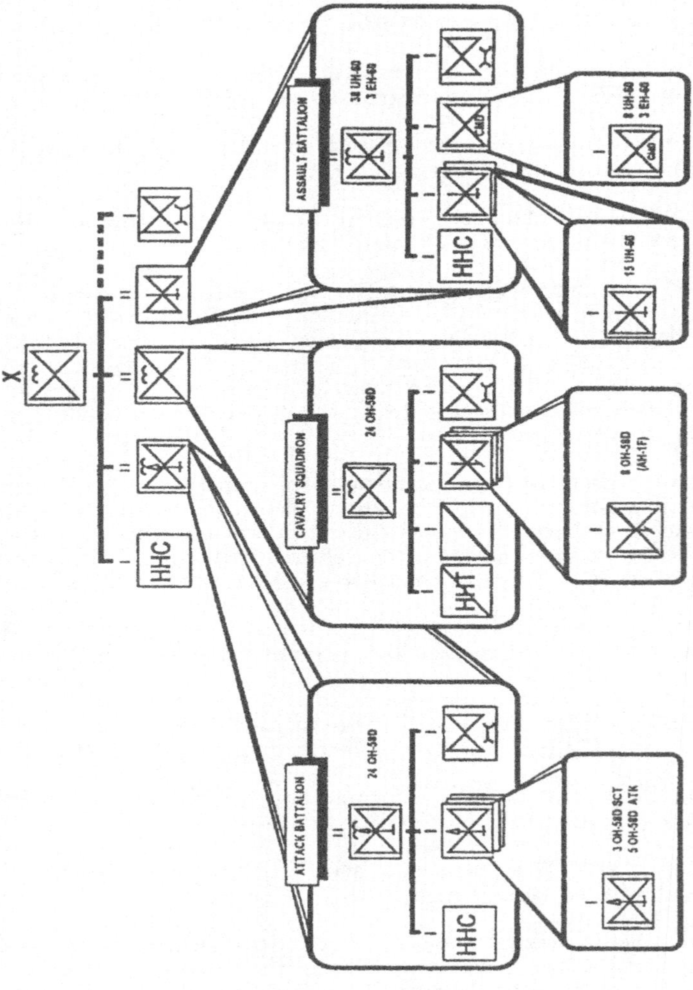

Figure A-5. Airborne division aviation brigade

that programs are established that reinforce collective training proficiency between the AHB and the units they are required to support.

(4) ***Air assault division aviation brigade.*** The primary mission of the air assault division aviation brigade (Figure A-6) is to deploy worldwide on short notice; plan, coordinate, and execute aviation operations as an integrated element of an air assault combined arms team; and find, fix, and destroy enemy forces in joint, combined, or unilateral operations. The air assault division aviation brigade is composed of a brigade HHC; one divisional cavalry squadron; one CH-47 Chinook helicopter battalion; three AHBs; one CAB; and three ATKHBs.

(a) The air assault division cavalry squadron consists of an HHT; an AVUM troop; and four ACTs. The four ACTs will be equipped with eight OH–58Ds each.

(b) The air assault division CH–47 Chinook helicopter battalion consists of an HHC and three CH–47 Chinook helicopter companies with 16 CH-47Ds in each company.

(c) The air assault division's three AHBs each consist of an HHC; an AWM company; and two assault helicopter companies (AHCs) with 15 UH–60s in each company. The limited assets of the support aviation company in the CAB will require the assault companies to perform utility missions as well.

(d) The air assault division CAB consists of an HHC; an AWM company; two CACS with eight UH–60s each; and a SAC with eight UH–60s and three EH–60s.

(e) The air assault division's three ATKHBs each consist of an HHC; an AWM company; and three ATKHCs equipped with eight AH-64s each.

(f) The DISCOM supports the aviation brigade with an AVIM battalion. The AVIM battalion has no organic aircraft.

(g) Utility helicopter support. The aviation brigade commander is the primary integrator of aviation within the division. The CAB supports the division as well as the aviation brigade. Utility aircraft support for the attack battalions, cavalry squadron, the CH–47 Chinook battalion, and the attached AVIM battalion will come from the CAB and the assault battalions. The brigade staff will routinely plan for utility support from the CAB and the assault battalions to the attack battalions, cavalry squadron and CH–47 Chinook battalion. In addition, the brigade commander will ensure that programs are established to reinforce collective training proficiency between the CAB and assault battalion and the units they are required to support.

A-4. ARMORED CAVALRY REGIMENTS

a. ***Armored cavalry regiment (heavy).*** The ACR is a self+ontained combined arms organization composed of armored cavalry squadrons (ACSs); an aviation squadron;

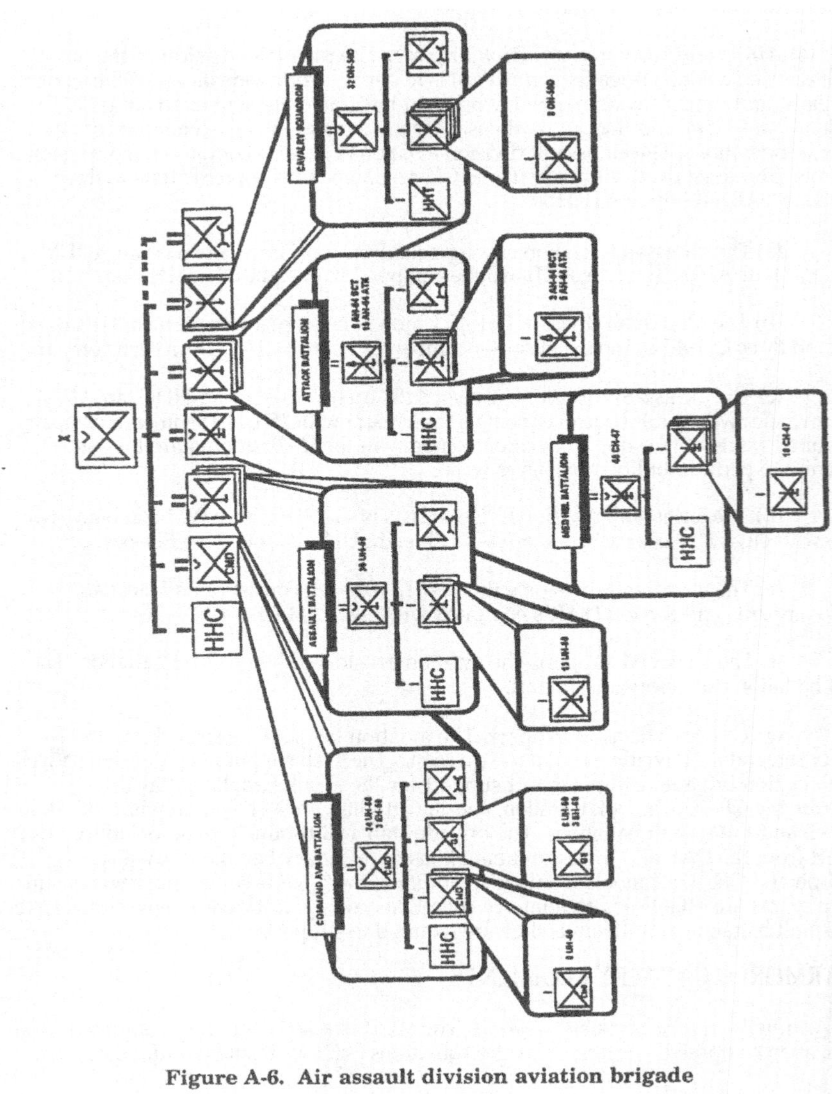

Figure A-6. Air assault division aviation brigade

a support squadron; and separate combat support companies and batteries. The ACR is a separate unit that supports the corps or a joint task force. It is often reinforced by corps combat support units and divisional maneuver battalions. The ACR operates independently over a wide area and at extended distances from other units. The ACR is a highly mobile armored force capable of fighting the fully mechanized threat in the environmental states of war or conflict. The ACR may be rapidly deployed to a theater of operations by sealift. When supporting a light corps, limitations may exist in corps support capabilities, strategic mobility, and terrain restrictions.

The regimental aviation squadron (RAS) (Figure A-7) provides the regiment with combat aviation assets. It is organized with air cavah-jdreconnaissance troops; attack helicopter troops; and an assault helicopter troop. The squadron adds a very responsive, terrain-independent movement capability to the regiment. Maneuverability and flexibility of the RAS enhance the combat flexibility of the regiment. The RAS may operate independently of, or in close coordination with, the ACS, or it may provide troops to the ACS.

b. ***Armored cavalry regiment (light).*** The ACR(L) is a self-contained combined arms organization capable of being packaged and rapidly deployed by air or sealift as part of a force projection Army responding rapidly to contingencies, worldwide. Its role may be traditional, initial entry, or follow-on. The traditional role would support a US corps or TF through a reconnaissance, security, and economy-of-force capability. As an initial entry force, the ACR(L) would support Army or joint task force operations with a credible force as a demonstration of US resolve. In the follow-on role, the ACR(L) will follow an opposed entry force (division–ready, brigade-type) to expand the point of entry; to provide reconnaissance and security, and to serve as the initial combat-capable maneuver force.

The RAS provides the regiment with combat aviation assets. It is organized with air cavalry troops equipped with eight OH–58D (Armed) helicopters each and a UH–60 Black Hawk assault helicopter troop. The squadron adds a very responsive, terrain-independent movement capability to the regiment. The maneuverability, firepower, and flexibility of the RAS enhance the combat flexibility of the regiment. The RAS may operate independently of, or in close coordination with, the ACS, or it may provide troops to the ACS.

A-5. CAVALRY SQUADRONS

a. ***Heavy division cavalry squadron.*** The heavy division cavalry squadron (Figure A-8) is a highly mobile, armor–protected force organized as part of the armored and mechanized infantry divisions. The squadron operates primarily in the environmental states of war and conflict. It may deploy into a theater by FAST sealift as part of a division, brigade, or joint task force. It consists of three M3/Ml-equipped ground troops and two OH–58D-equipped air cavalry troops. The squadron can expect to perform reconnaissance and security operations in conditions characterized as fluid and continuous, mobile, wide ranging, and firepower intensive.

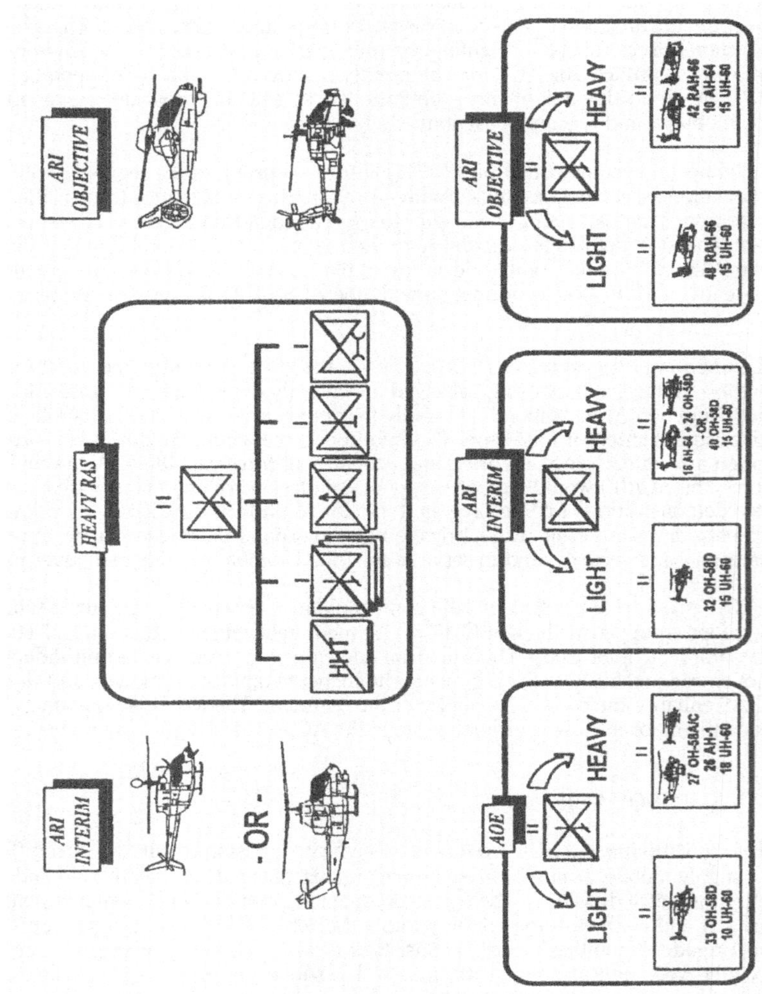

Figure A-7. Regimental aviation squadron

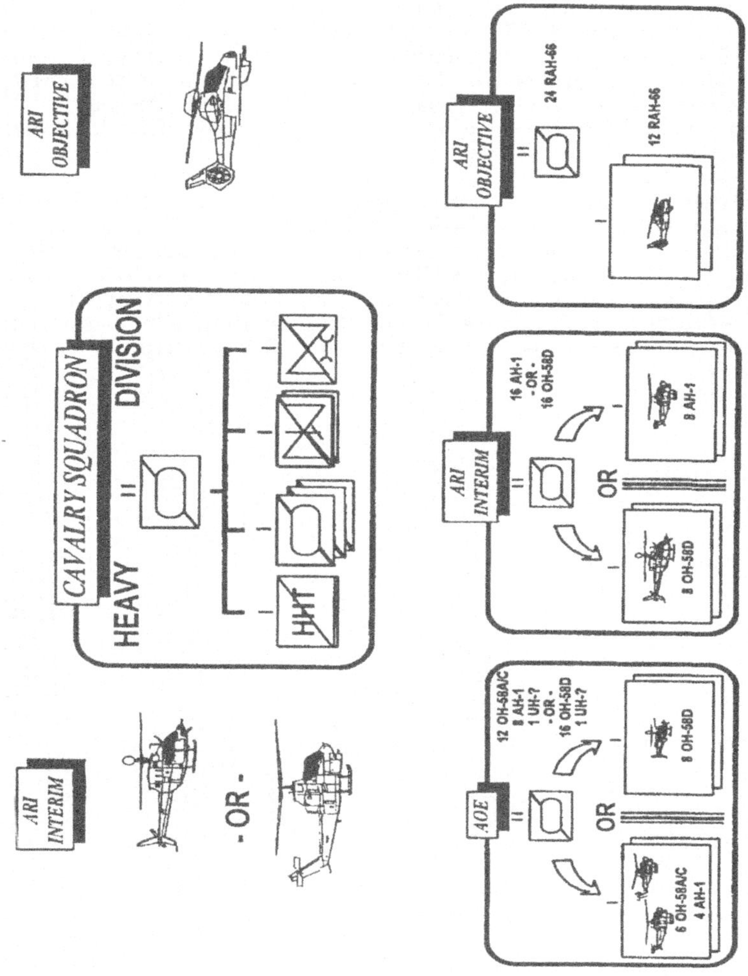

Figure A-8. Heavy division cavalry squadron

b. ***Light division cavalry squadron.*** The light division cavalry squadron (Figure A-9) is a high–mobility, lightly armed force organized as part of light infantry divisions. It consists of one high–mobility, multipurpose wheeled vehicle (HMMWV)-equipped ground troop and and two OH–58D (Armed) helicopter-equipped air cavalry troops. As part of the light infantry division, this squadron may operate in any environmental state from peace to war. The squadron is deployable by air or sealift to a theater of operations as part of the division, or in support of a brigade or joint task force. This squadron possesses a significant tactical mobility advantage over the infantry battalions in the division.

c. ***Air cavalry squadron.*** The air cavalry squadron is a highly mobile, armed force organized as part of airborne (Figure A-10) and air assault (Figure A-11) divisions. It is also organic to those corps without an assigned ACR. It is equipped predominantly with air cavalry troops, each containing eight OH–58D aircraft. Ground cavalry troops may be part of the squadron. When present, ground troops are mounted in HMMWVs. The squadron is structured light to possess the same strategic mobility as the parent division. When deployed, the squadron possesses a significant mobility advantage over infantry battalions of the division.

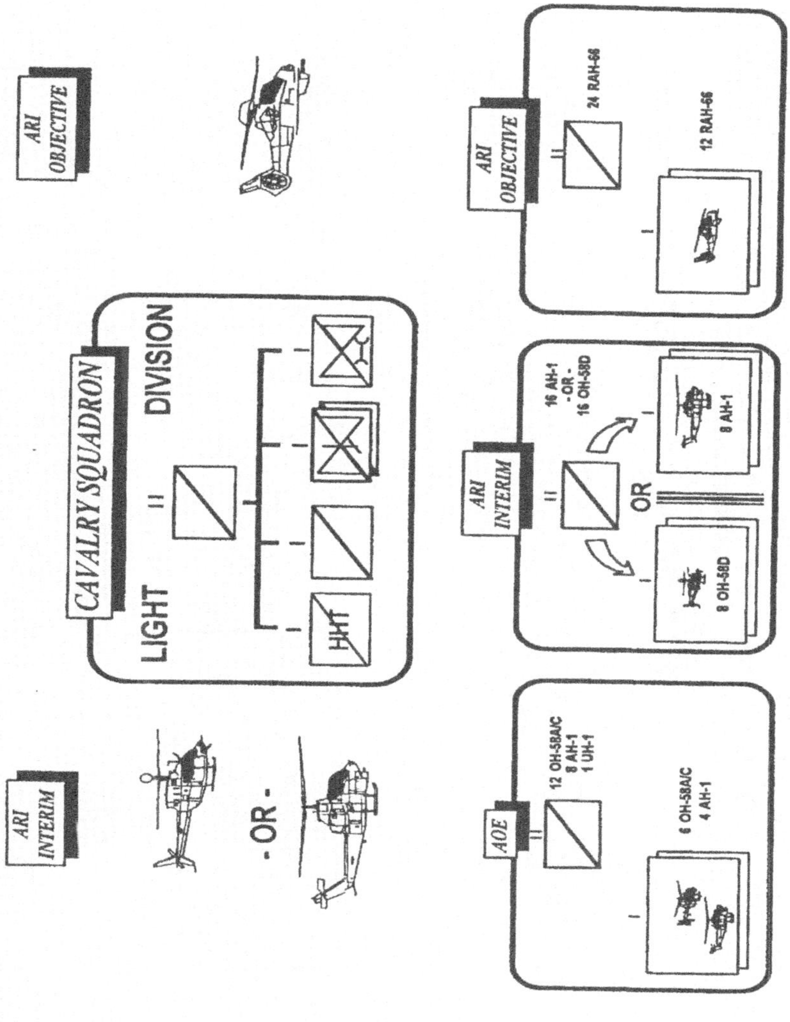

Figure A-9. Light division cavalry squadron

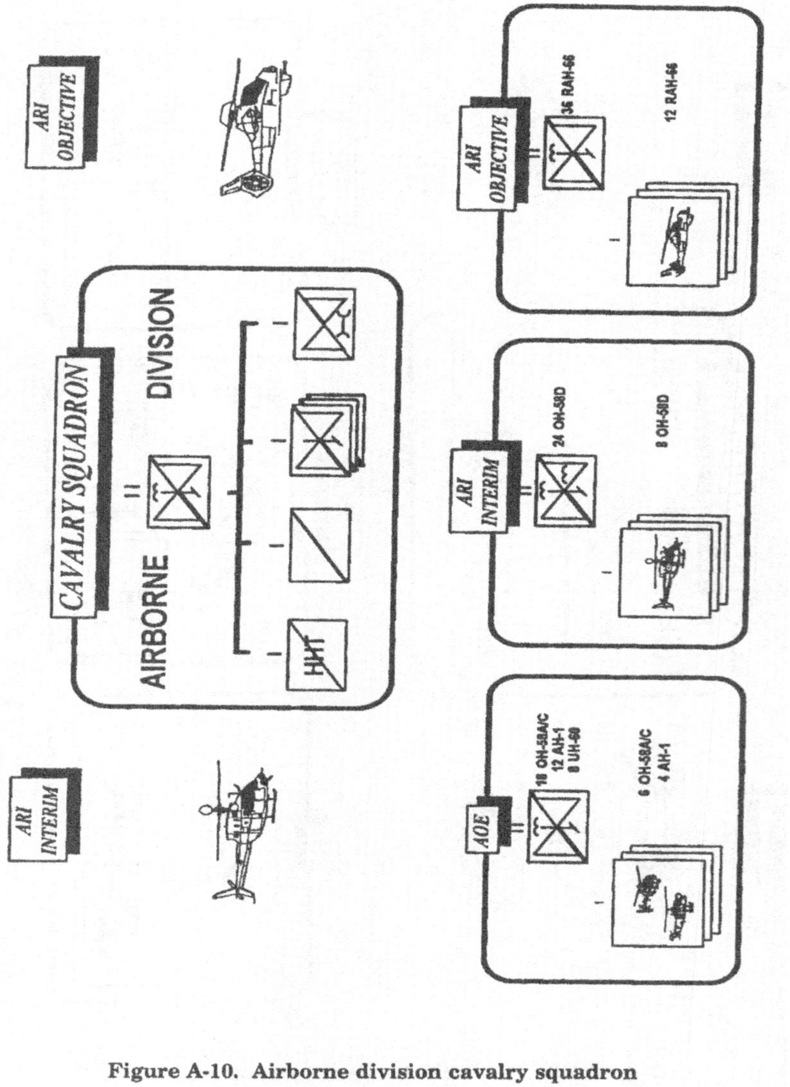

Figure A-10. Airborne division cavalry squadron

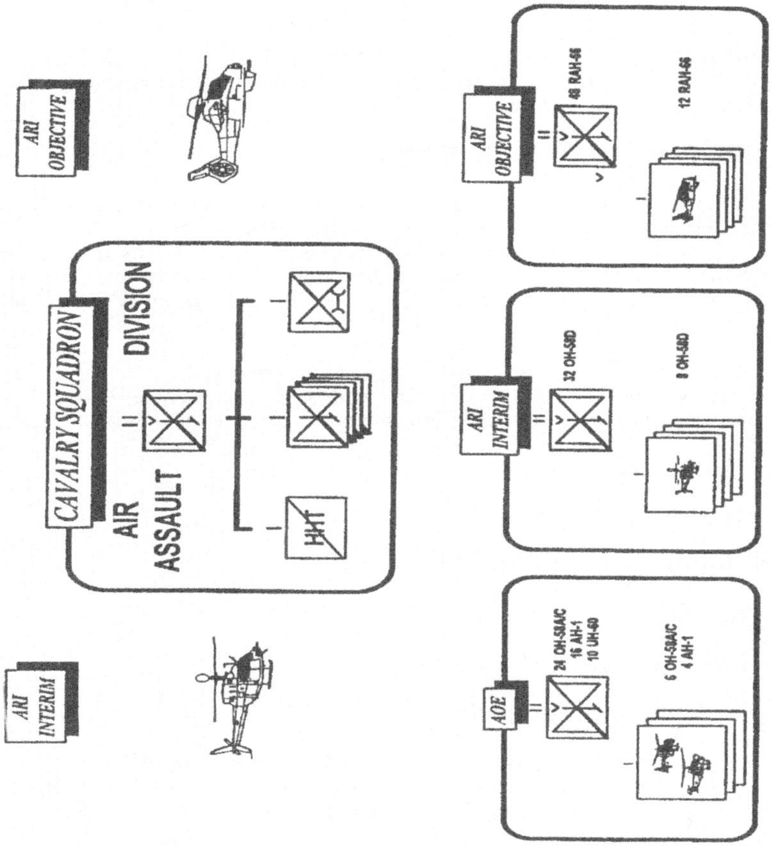

Figure A-11. Air assault division cavalry squadron

APPENDIX B

AVIATION COMMAND AND CONTROL

Command and control (C^2) is vital to synchronized Army operations. This appendix discusses those elements essential to aviation C^2. Emerging G systems are discussed in appendix D.

B-1. COMMAND AND CONTROL RESPONSIBILITY

a. Coordination of Army aviation C^2 begins with the force commander; he alone is responsible for the outcome of combat actions on the battlefield. Army aviation commands and controls in the same manner as any other combined arms organization. However, battlefield distances over which aviation units routinely operate encompass the entire area of operations (AO) of the force. This makes the commander's decision about where to place himself very difficult.

b. Thus, special considerations must be given to ensure effective C^2. Army aviation has certain C^2 requisites that support the force with agile air maneuver. These requisites, which can synchronize massed aviation combat power against enemy weaknesses, are—

- Timely intelligence.
- Reliable communications.
- Effective aviation liaison.
- Accurate weather forecasting.
- Flexible, mobile command posts (CPs).
- Efficient airspace coordination.

c. The aviation commander must advise the force commander about the C^2 capabilities provided by Army aviation when employed with combined arms, joint, combined, or special operations forces. Army aviation commanders from echelons above corps (EAC) to battalion level are responsible for coordinating, through the chain of command, with the force commander and assisting him in the employment of Army aviation forces.

d. Another critical element of Army aviation C^2 is the C^2–protect function. The purpose of C^2–protection is to maintain friendly capabilities for effective battle command by negating or turning to friendly advantage adversary counter–C^2 actions. These actions can be both active and passive. Passive means include operational security (OPSEC); communications security (COMSEC); and survivability measures like mobility, hardening, and geographic dispersal to limit our exposure and vulnerability to electronic warfare (EW) attacks. Active means involve both the direct attack of adversary counter–C^2 resources and actions taken to limit his ability to plan and control the counter–C^2 effort. FM 100–6 contains a detailed discussion of the C^2 protect function.

B-2. COMMAND POSTS

a. Aviation uses highly mobile ground and aerial command posts to control and support combat operations. Army aviation's C^2 requirements are similar to those of any other maneuver force. Aviation headquarters normally establish a main and rear CP. When required, a tactical CP is established. This CP distribution allows aviation to operate efficiently; it minimizes the difficulty of establishing and sustaining more than two echelons of control at any one time.

b. The main CP controls current aviation operations while planning future operations; it is normally located beyond the effective range of threat artillery.

c. The rear CP concentrates on sustaining the aviation force. It should be located near the force logistics base and elements. When required, a forward tactical CP will locate forward to direct and provide control for high–tempo aviation maneuver operations. A tactical CP usually is configured with a limited number of mobile ground assets; it may also be in an aerial platform.

d. Aviation commanders at all levels position themselves to best influence the conduct and outcome of aviation and combined arms operations. This is the most critical task. They must be prepared to accept and C^2 ground forces for limited periods of time. Normally, such operations are mission specific. However, objectives assigned must equate to the combat power and sustainment capability of an organized task force. Figure B–1 illustrates a typical brigade C^2 network.

B-3. COMMUNICATIONS

a. Army aviation commanders and staffs need reliable, long–range, redundant communication systems. With these, they can exercise effective C^2 throughout the supported force's AO. Aviation forces must maintain and sustain the same communications capabilities as any other maneuver force.

b. Radio is normally the primary means of internal and external communications. To expediate C^2, Army aviation uses the following: frequency modulated (FM), high–frequency (HF) amplitude modulated (AM) voice, ultra–high frequency (UHF), very high frequency (VHF), common–user systems, and internal wire.

c. When operating over extended battlefield distances, aviation commanders must have access to varying battlefield communication systems for relay and retransmission capabilities. These systems help ensure uninterrupted battlefield communications, especially at terrain flight/nap–of-the-earth (NOE) altitudes.

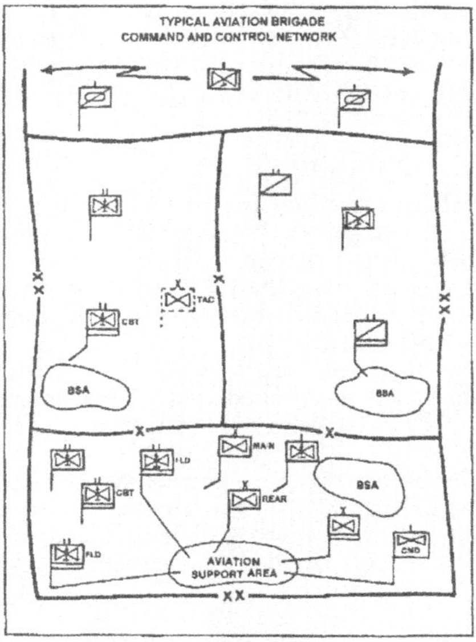

Figure B-1

B-4. INTELLIGENCE

a. Army aviation's combat success depends on the aviation commander's ability to *"see the battlefield."* Accurate and timely intelligence about the enemy, weather, and terrain is imperative for all aviation operations. Because of the density, lethality, and sophistication of threat target acquisition radars and air defense (AD) weapon systems, Army aviation must have current intelligence information that encompasses the supported maneuver force AO.

b. To see the battlefield, the aviation commander must drive his intelligence battlefield operating system (BOS) by specifying his priority intelligence requirements (PIRs) and targeting needs. He requires his intelligence assets to provide the intelligence he needs in the form he can use in time to support his decision-making process. Aviation commanders must be able to create and take advantage of windows of opportunity. They also must be able to exploit enemy vulnerabilities and weaknesses. They must know—

- The entire force AO.

- The current and forecasted environmental conditions.

- The terrain and air avenues of approach.
- The nature, capabilities, and activities of the threat.

B-5. AIRSPACE COORDINATION

a. Airspace coordination is critical to all Army aviation operations. Army airspace command and control (A^2C^2) is the Army's process for accomplishing airspace coordination. Potential users of the aerial dimension of the battlefield include Army aviation, AD, military intelligence, fire support, and jointicombined air forces.

b. A^2C^2 maximizes joint effectiveness by ensuring simultaneous airspace use-synchronized in time, space, and purpose—to produce maximum combat power at decisive moments. The maneuver commander's plans must address the best use of the airspace throughout the AO. Using airspace effectively enhances total operational capabilities and is a key node in fratricide prevention.

c. Another key element in fratricide prevention is the correct use of published identification, friend or foe (radar) (IFF) modes and codes. The airspace control authority directs the tactical use of IFF in each theater. Commanders must ensure that the IFF equipment in their aircraft is always coded, maintained, and used properly to ensure the safety of their flight crews.

B-6. DEEP OPERATIONS COORDINATION CELL (DOCC)

a. The DOCC now exists at every corps and many division tactical operations centers to facilitate the conduct of deep operations. The DOCC allows corps and division staffs to coordinate all aspects of the deep battle at a single location. This coordination directly benefits an aviation commander.

b. Functions of the DOCC include—

- Planning and executing deep operations in support of corps/division operational plans (OPLANS).
- Synchronizing combat, combat support (CS), and combat service support (CSS) to support deep operations.
- Determining high–payoff targets for deep operations.
- Interfacing with the Joint Targeting Coordination Board (JTCB) and the corps targeting cell to provide a linkage between joint and organic fires.
- Developing the detection and delivery concepts to support deep operations.

APPENDIX C

US ARMY AIRCRAFT CAPABILITIES

Table C-1. Fixed wing aircraft capabilities

A. AIRCRAFT[1]	UNIT	C-12C	C-12D	RC-12D	RV-1D*
B. NORMAL CREW	per aircraft	2	2	2	2(Pilot & Operator)
C. OPERATIONAL CHARACTERISTICS[2,3,4]					
(1) Maximum allowable gross weight	lbs	12,500	12,500[23]	14,200	18,109
(2) Basic weight	lbs	8, 084	8,084	8,143	12,054
(3) Useful load	lbs	4,416	4,416	2,078	6,055
(4) Payload/Normal mission	lbs	2,000	2,000	2,000	NA
(5) Fuel capacity[9] (Internal/external)	lbs/gal	2,470/386	2,470/386	2,470/386	1,790/276 1,950/300
(6) Fuel consumption rate[9]	lbs/gal/hr	456/70	456/70	456/70	900/130
(7) Normal cruise	kts	260	260	260	220
(8) Endurance at cruise (Plus 30 + min reserve)	hrs + minutes	5 + 15	5 + 15	5 + 15	1 + 40[14] 3 + 30[15]
(9) Grade of fuel	octane	JP-4/5/8	JP-4/5/8	JP-4/5/8	JP-4/8
D. PASSENGER CAPACITY					
(1) Troop seats	ea	8	8	8	1
(2) Normal capacity	ea	8	8	8	1
(3) Total capacity w/crew	ea	10	10	10	2
(4) Litters & ambulatory	ea	NA	NA	NA	NA
E. EXTERNAL CARGO					
(1) Maximum recommended external load[5]	lbs	NA	NA	NA	ea wing 2,000

Table C-1. Fixed wing aircraft capabilities—continued

A. AIRCRAFT[1]	UNIT	C-12C	C-12D	RC-12D	RV-1D*
F. DIMENSIONS					
(1) Length—fuselage[8]	ft-in	43'10"	43'10"	43'10"	41'9"
(5) Width—tread	ft-in	17'2"	17'2"	17'2"	9'2"
(6) Height—extreme	ft-in	15'5"	14'9"	15'5"	13'0"
(9) Wing span	ft-in	54'6"	55'6.5"	54'6"	48'0"
G. CARGO DOOR					
(1) Dimensions—width/height	in	27.7" x 51.5"	52" x 52"	27.7" x 51.5"	NA
(2) Location—side of fuselage	left/right front/rear	left rear	left rear	left rear	NA
H. CARGO COMPARTMENT					
(1) Floor—above ground	in	47	42	47	NA
(2) Usable length	in	128	128	128	NA
(3) Floor width	in	54	54	54	NA
(4) Height (clear of obstructions)	in	57	57	57	NA
(5) Maximum cargo space	cu ft	306.5	306.5	30	

Table C-2. Rotary wing aircraft capabilities

A. AIRCRAFT[1]	UNIT	OH-6A	OH-58A	OH-58C	CH-47D	OH-58D
B. NORMAL CREW	per acft	1+observer	1+observer	1+observer	4	2
C. OPERATIONAL CHARACTERISTICS[2,3,4]						
(1) Maximum allowable gross weight	lbs	2,400	3,000	3,200	50,000	5,200
(2) Basic weight	lbs	1,163	1,586	1,898	22,499	
(3) Useful load	lbs	1,237	1,417	1,302	27,501	
(4) Payload/Normal mission	lbs	650[8]	760[8]	837[8]	20,206	
(5) Fuel capacity[9] (inter/ext)	lbs/gal	400/61.5	475/73	465/71.5	6,695/1030	762/112
(6) Fuel consumption rate[9] (internal/external)	lbs/gal/hr	143/22	189/29	175/27	2,600/400	280/41
(7) Normal cruise	kts	90	90	90	120	90-100
(8) Endurance at cruise (Plus 30 + min reserve)	hrs + minutes	3 + 15	3 + 30	3 + 00	2 + 30	2 + 00
(9) Grade of fuel	octane	JP-4/5/8	JP-4/5/8	JP-4/5/8	JP-4/5/8	JP-4/5/8
D. PASSENGER CAPACITY						
(1) Troop seats	ea	3	2	2	33	0
(2) Normal capacity	ea	3	4	4	33	0
(3) Total capacity w/crew	ea	4	4	4	37	0
(4) Litters & ambulatory	ea	NA	2/4	2	24	0
E. EXTERNAL CARGO						
(1) Maximum recommended external load[5]	lbs	NA	NA	NA	26,000	NA
(2) Rescue hoist capability	lbs	NA	NA	NA	600	NA
(3) Cargo winch capacity	lbs	NA	NA	NA	3,000	NA
F. DIMENSIONS						
(1) Length—fuselage[6]	ft-in	23'0"	32'3.5"	32'8.8"	51'0"	40'

Table C-2. Rotary wing aircraft capabilities—continued

A. AIRCRAFT[1]	UNIT	OH-6A	OH-58A	OH-58C	CH-47D	OH-58D
(2) Length—blades unfolded	ft-in	30'4"	40'11.8"	40'11.8"	99'0"	40'6.4"
(3) Length—blades folded	ft-in	23'0"	NA	NA	51'0"	40'
(4) Width—blades folded	ft-in	5'6"	NA	NA	12'5"	11'
(5) Width—tread	ft-in	6'9"	6'3.5"	6'5.4"	11'11"	6'5.4"
(6) Height extreme	ft-in	8'3"	9'6.5'	12'0"	18'8'	12'10.6"
(7) Diameter—main or forward rotor	ft-in	26'4"	35'4"	35'4"	60'0"	35'
(8) Diameter—tail or rotor	ft-in	4'3"	5'2"	5'2"	60'0"	5'5"
(9) Wing span	ft-in	NA	NA	NA	NA	NA
G. CARGO DOOR						
(1) Dimensions/height	in	41"x34.5"	40"x35"	40"x35"	90"x78"	NA
(2) Location—side of fuselage (left & right/front and rear		left & right	left & right	rear right	rear	
H. CARGO COMPARTMENT						
(1) Floor—above ground	in	14.5"	22.5"	22.5"	31.2"	NA
(2) Usable length	in	5'9"	39"	39"	30'2"	
(3) Floor width	in	3'2"	50"	50"	7'6"	
(4) Height (clear of obstructions)	in	3'2"	50"	50"	6'6"	
(5) Maximum cargo space	cu ft	40	20	20	1,474	
I. Weapons[21]	NA	NA	NA	NA	M60	ATAS HELLFIRE .50 2.75

Table C-2. Rotary wing aircraft capabilities—continued

A. AIRCRAFT[1]	UNIT	AH-1F*	AH-64[13]	UH-IHV[22]	UH-60A/L[23]
B. NORMAL CREW	per acft	2	2	2	3
C. OPERATIONAL CHARACTERISTICS[2,3,4]	lbs				
(1) Max allowable gross weight	lbs	10,000	17,400	9,500	20,000
(2) Basic weight	lbs	6,598	10,505	5,132	10,500
(3) Useful load	lbs	4,302	6,895	4,368	7,945
(4) Payload/Normal mission	lbs	1,293[20]	4,090[20]	2,900	3,360[16]
(5) Fuel capacity[9] (Internal/external)	lbs/gal/hr	1,703/262	2,405/370	1,358/209	2,360/362
(6) Fuel consumption rate[9]	lbs/gal	640/98	810/124	550/84	960/148
(7) Normal cruise speed	kts	90-130[19]	90-140[19]	90-120	120-190
(8) Endurance at cruise (Plus 30 min reserve)	hrs + mins	2 + 30	1 + 45	2 + 15	2 + 15
(9) Grade of fuel	octane	JP-4	JP-4/5/8	JP-4/5	JP-4/5/8
D. PASSENGER CAPACITY					
(1) Troops seats	ea	0	0	11	14
(2) Normal capacity	ea	0	0	11	14
(3) Total capacity w/crew	ea	0	0	13	17
(4) Litters and ambulatory	ea	0	0	6	4/6
E. EXTERNAL CARGO					
(1) Maximum recommended external load[5]	lbs	1,380[18]	6,200[18]	4,000	8,000
(2) Rescue hoist capacity	lbs	NA	NA	300[17]	600
(3) Cargo winch capacity	lbs	NA	NA	NA	NA
F. DIMENSIONS					
(1) Length—fuselage[6]	ft-in	44'7"	49'3"	40'7"	50'7.5"

Table C-2. Rotary wing aircraft capabilities—continued

A. AIRCRAFT[1]	UNIT	AH-1F	AH-64[13]	UH-1H/V[22]	UH-60A/L[23]
(2) Length—blades unfolded	ft-in	53'1"	57'1"	57'1"	64'10"
(3) Length—blades folded	ft-in	NA	NA	NA	40'4"
(4) Width—blades folded	ft-in	10'9"	16'3"	8'7"	9'8.1"
(5) Width—tread	ft-in	7'0"	6'6"	8'7"	8'10.2"
(6) Height—extreme	ft-in	13'9"	12'6"	14'6"	17'6"
(7) Diameter—main or forward motor	ft-in	44'0"	49'0"	48'	53'8"
(8) Diameter—tail or rear rotor	ft-in	8'6"	9'3"	8'6"	11'0"
(9) Wing span	ft-in	10'4"	16'3"	NA	NA
G. CARGO DOOR					
(1) Dimensions—width/height	in	NA	NA	74"x48"	68"x54"
(2) Location—side of fuselage	left & right front rear	NA NA	NA NA	left & right	left & right
H. CARGO COMPARTMENT					
(1) Floor—above ground	in	NA	NA	24	19
(2) Usable length	in	NA	NA	92	110
(3) Floor width	in	NA	NA	96	72
(4) Height (clear of obstructions)	in	NA	NA	49	54
(5) Maximum cargo space	cu ft	NA	NA	220	246.8
I. WEAPONS[10]	NA	TOW 20mm 2.75	30mm HELLFIRE 2.75	M60	M60

Table C-3. Special operations aircraft capabilities

A. AIRCRAFT[1]	MH-60K	MH-47E	C-23
B. NORMAL CREW	4	5	3
C. OPERATIONAL CHARACTERISTICS[2,3,4]			
(1) Max allowable gross weight (lbs)	24,500	54,000	25,600
(2) Basic weight	14,265	26,918	16,874
(3) Useful load	10,235	27,082	7,280
(4) Payload normal mission	5,000	N/A	N/A
(5) Fuel capacity (lbs)	2,340	13,442	4,355
(6) Fuel consumption (lbs/hr)	960	2,600	1,000
(7) Normal cruise speed (kts)	120	140	180
(8) Endurance at cruise speed (Plus 30 + min reserve)	2 + 15	4 + 36	4 + 21
(9) Grade of fuel	JP-4/5/8	JP-4/5/8	JP-4/5/8
D. PASSENGER CAPACITY			
(1) Troop seats	14	44	30
(2) Normal capacity (ea)	14	44	30
(3) Total capacity (ea)	18	49	33
(4) Litters & ambulatory	N/A	24/44	15/3
E. EXTERNAL CARGO			
(1) Maximum recommended external load (lbs)	8,000	26,000	NA
(2) Rescue hoist capability (lbs)	600	600	NA
(3) Cargo winch capability (lbs)	NA	3,000	NA

Table C-3. Special operations aircraft capabilities—continued

A. AIRCRAFT[1]	MH-60K	MH-47E	C-23
F. DIMENSIONS			
(1) Length—fuselage (ft)	50.7	51.0	58.04
(2) Width—tread	9.0	11.1	13.9
(3) Height—extreme	17.8	18.8	16.41
(4) Length—blades unfolded	64.1	99.0	N/A
(5) Length—blades folded	57.4	51.0	N/A
(6) Width—blades folded	14.4	12.5	N/A
(7) Wing span	N/A	N/A	74.84
(8) Diameter—main rotor(s)	53.8	60.0	N/A
(9) Diameter—tail rotor	11.0		
G. CARGO DOOR			
(1) Width (in)	68.0	90.0	65.0
(2) Height (in)	54.0	78.0	55.0
(3) Location on fuselage	left/right	rear	left
H. CARGO COMPARTMENT			
(1) Floor above ground (in)	19.0	31.2	10.0
(2) Usable length (ft)	9.1	30.2	29.82
(3) Floor width (ft)		7.6	6.2
(4) Height (ft)	4.5	6.6	6.2
(5) Maximum cargo (cu ft)		1,474	1,246
I. WEAPONS (miniguns)	2 x M-134	2 x M-134	

Tables C-1 through C-3. US Army aircraft capabilities-continued

Footnotes:

[1]A - Attack, C - Cargo, O - Observation, U - Utility.
[2]All data computed at standard conditions at sea level.
[3]Detailed weight computations and characteristics taken from current 55-series TMs.
[4]Data subject to change due to developmental testing.
[5]Maximum load the aircraft is capable of lifting.
[6]Dimension from nose to end of tail.
[7]Varies with load carried. Figure given is for normal mission profile.
[8]Does not meet 200-nm range requirement of normal mission definition.
[9]Aviation gas figured on 6 lbs/gal. JP-4 computed on 6.5 lbs/gal. JP-8 computed on 6.8 lbs/gal.
[10]Indicates type of weapons aircraft can carry. Specific armament based on unit assignment.
[11]Seven-round 2.75-inch rocket pod.
[12]Nineteen-round 2.75-inch rocket pod.
[13]Subject to final development configuration.
[14]Without external fuel.
[15]With external fuel.
[16]Normal mission, internal load, probability exists to cube out before weight out. Max load on the floor is 300 lbs/sq ft.
[17]UH-1 is restricted to hoist capacity of 300 lbs because of center of gravity conditions.
[18]External wing stores.
[19]Due to armament configurations and flight profiles.
[20]Considers gross weight minus basic weight minus 400 lbs for crew and total fuel weight.
[21]Fuel consumption at 92 kts, 77 gal/hr; at 140 kts, 84 gal/hr.
[22]Weapons are not applicable to UH-1V, medical evacuation helicopters.
[23]MWO allows 13,000 or 13,500 Ibs per specified airframe.
[24]Reserve/NG only.

Considerations:

- These tables are for general reference use only. Refer to the appropriate operator's manual for detailed information. Definitions of terms used in this table include the following:
- *Maxknmn allowable gross weight.* The maximum allowed total weight of the aircraft before takeoff. The "basic weight" of the aircraft plus the crew, personnel equipment, special device, passengers and cargo, and usable fuel and oil. This is limited by structure, power available, or landing load.
- *Basic weight.* The empty weight of an aircraft configuration to include all appointments, integral equipment, instrumentation, trapped fuel and oil, but excluding passengers, cargo, crew, fuel, and oil.
- *Useful load.* The load-carrying capability of an aircraft including payload, crew, oil, and usable fuel required for the mission. This is the difference between "maximum allowable gross weight" and "basic weight" as defined above. Thus, a reduction of the fuel load will decrease the endurance and increase the payload. Fuel oil is required for all missions.
- *Payload.* The useful load less the crew, full oil, and the required fuel for the mission.
- *Normal mission.* Payload available computed under the following conditions:
  - Fuel for 200 nm plus a 30-minute reserve.
  - Flight altitude 2000' mean sea level, standard temperature.
  - Takeoff maximum gross weight (weight of crew included).
- *Normal cruising speed.* The true airspeed that an aircraft can normally be expected to maintain at some standard power setting below rated military power. This speed will vary with altitude; for example, the U-8F's normal cruising speed is 165 at 65 percent power at 8,000 ft.
- *Endurance at cruising speed.* The time that an aircraft can remain airborne at normal cruising speed with fuel aboard without using the required fuel reserve. The data listed under "Operational Characteristics" are computed using full fuel minus a 30-minute reserve.

APPENDIX D

DIGITIZATION

"Never forget that all technology can ultimately do is give your staff more time to think. It can't think for them. Data is not information. Information is not judgment. Judgment is not wisdom. Numbers aren't policy. Quantitative approaches can't solve qualitative problems."

Joshua Shapiro, Technology Consultant

Section I. INFORMATION TECHNOLOGY

D-1. INTRODUCTION

a. This appendix reflects and supports the doctrine for future Army aviation operations. It emphasizes the importance of synchronization of the digitized aviation task force (TF) through the real–time exchange of battlefield digital information. This rapid exchange of information will be accomplished through—

- Digital intelligence information.
- Digital mission planning and loading.
- Digital situational information.
- Synchronous command and control (C^2) of the battlefield operating systems.

b. The aviation TF commander and his staff will use this information to rapidly assess the tactical situation, maneuver to the decisive point on the battlefield, and destroy the enemy. To date, all digital experiments have confirmed that our doctrine is sound; they have shown that digital equipment allows us to plan, prepare, and execute better—but not necessarily faster. We have only scratched the surface on digitization, and the future holds great promise.

D-2. DEFINITION, GOAL, AND PURPOSE

a. A brief definition of digitization follows: *Digitization* is the application of information technology to acquire, exchange, and employ timely digital information tailored to the needs of each decider, shooter, and supporter, allowing a clear and accurate vision of the battlespace.

b. The Army's goal is to introduce modern information technology throughout the force in a way that optimizes capability. Today, the Army organizes units around weapons systems —*"feeding the guns."* In the near future, the Army will organize units to maximize information; that is, to create and share knowledge. Unified action will follow based on that knowledge, which will allow commanders to apply power effectively.

c. Our purpose will be to dominate, to control, and to win; information will be the means to a more powerful end. Information–based battle command will give the Army ascendancy and freedom of action for decisive results in 21st century warfare and peace operations.

D-3. ENHANCED CAPABILITIES

a. The digitally equipped aviation TF possesses significantly enhanced capabilities over the nondigitally equipped force. It electronically links air and ground forces allowing the commander to synchronize combat power with devastating effects. This integration is accomplished by the following seven aviation modernization programs—

- Global Positioning System (GPS).
- HaveQuick II (HQ II) UHF–AM radio.
- High Frequency Nap-of-the-Earth Communications (HF NOE COMM).
- Improved Data Modem (IDM).
- Aviation Mission Planning System (AMPS).
- Army Airborne Command and Control System (A^2C^2S).
- Aviation Tactical Operations Center (AVTOC).

b. What differentiates the digitally equipped force is the scope, intensity, and tempo of contemporary versus future operations brought on by the timeliness and accuracy of information provided by information age systems and sensors. Aviation information age systems reside at critical battle nodes. The Army's aviation operational architecture is composed of three information nodes. They are the AVTOC, A^2C^2S, and the collective fleet of modernized aircraft (OH–58D Kiowa Warrior, AH-64 Longbow Apache, and RAH-66 Comanche).

c. The AVTOC serves as the planning and primary synchronization point for the entire aviation operation. A^2C^2S focuses on the execution of current operations. It permits command and control on–the–move allowing the commander to influence action throughout the battlespace. The collection fleet of modernized aircraft uses data and imagery to conduct target acquisition and direct precision fires. Their sensors also create battlefield information that can be shared with commanders, other weapons systems, and intelligence providers.

d. The digitally equipped TF (Figure D-1) provides the commander—

- Increased situational awareness and sensor–to-shooter links.
- Enhancements to the mission planning process, orders preparation, intelligence flow, and distribution process.
- Digital aids that enhance the timeliness of tactical decisionmaking.
- Synchronized fire support.

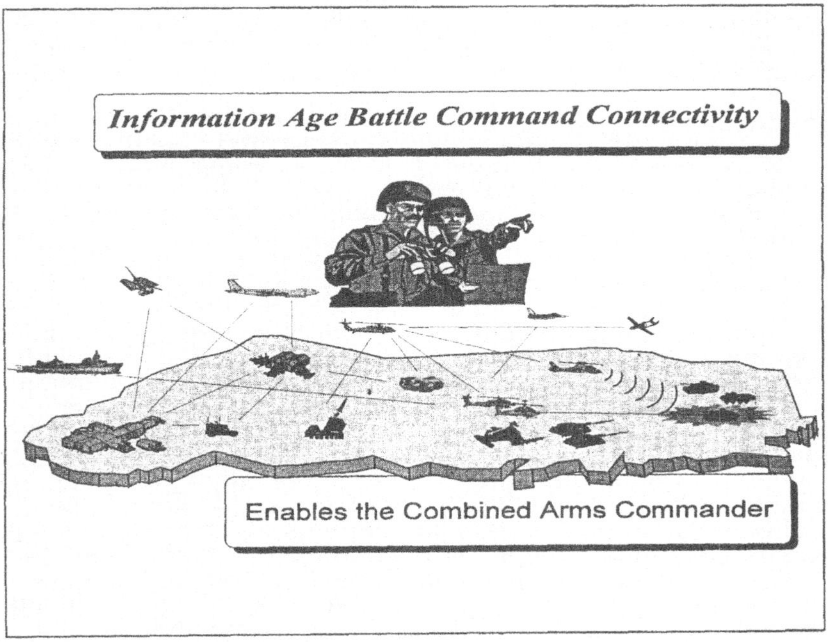

Figure D-1

D-4. SITUATIONAL AWARENESS

The increased situational awareness that the following systems and other digital systems provide the commander significantly enhances C^2: A^2C^2S, the Aviation Tactical Operations Center (AVTOC), the Enhanced Position Location Reporting System (EPLRS), and the AMPS. By observing the displays on these systems, the commander can see situational depictions of both the friendly and threat forces. With the aid of the GPS in the cockpit, the commander can instantaneously now know his precise location.

D-5. PUNNING PROCESS

a. The AVTOC—equipped with digital systems such as All Purpose Analysis System–Warrior (ASAS–W) and AMPS-provides significant enhancements to aviation mission planning and execution. The aviation TF commander and staff can issue warning orders with draft operations overlays early in the orders development process. This enhancement should allow subordinate commanders to begin their troop–leading procedures, rehearsals, and reconnaissance much earlier than previously possible. Digital information is exact. All subordinate elements receive exactly what is sent by higher headquarters. This reduces many human errors such as transposing grid coordinates.

b. When the staff has completed the planning process, and is ready to issue the operation order (OPORD), they digitally update and send the operational graphics before the issuance of the order. This allows subordinate units to initiate additional detailed planning while the commander is away receiving the OPORD. Mission, enemy, terrain, troops, and time available (METT–T) and creativity determine how a unit can most effectively use the digital systems available to them.

D-6. REPORTING

a. The ability to send tactical reports digitally helps the commander to shape the battlefield and react to the changing tactical situation. Pre-formatted reports via the improved data modem (IDM) allow the commander to assess the information reported. Spot reports (SPOTREPs), situation reports (SITREPs), and battle damage assessment (BDA) reports are just a few of the digital reports. These reports can be sent *"broadcast"* to everyone on the net or *"addressed"* to only those with a need to know.

b. Grid locations in the reports are more accurate since they are obtained from digital sources, i.e., GPS or locating laser targets. Routine reports can be formatted before the mission to reduce the time it takes to compile the report in flight. Sending digital imagery or graphics with a report may enhance the quality of the report. For example, an OH–58D pilot sends a BDA report with text and a digital image of the degree of destruction; an AH-64D pilot sends graphics files from his radar sweep to subordinate team members to assign sectors of fire.

D-7. LIMITATIONS

Digital systems and their use do have limitations despite their numerous advantages. Some of these limitations can have significant effects. A good backup plan should always be rehearsed. Some of these limitations include, but are not limited to, the following:

1. Some key elements of the task force, such as CS or CSS units, may not be equipped with digital systems.

Mixing digital and nondigital units in the same TF presents a special challenge to the TF commander and staff. Procedures must be established to communicate and share information. Control measures used on digital overlays must match those on conventional hard copy overlays.

2. Physical limitations of the hardware or software.

Each version of software may have different capabilities. For example, one version of AMPS may allow only a certain number of waypoints to be entered. Limitations of the hardware may include memory capacities, processing speed, or printed output. Large graphics files may slow down digital transmissions to the point at which they may become counterproductive. In some cases, voice reporting may be more expedient and digital reports are used for follow–up reports. The commander must decide when digital is better. Digital communications links may be interrupted.

3. Digital communications nets require precise procedures and strict net discipline.

Combining digital, voice, and imagery/graphics transmissions on the same net requires strict discipline. If communications net allocations permit, separate nets for voice and digital may be desired. Frequency hopping radio setup in HaveQuick II (HQ II) and single channel ground and airborne radio system (SINCGARS) requires precise entries. Use of a data loader or electronic fill reduces setup errors. Using GPS to time-frequency hopping is the preferred method to synchronize the net.

4. Digital to nondigital information/data exchange (and vice versa) requires additional time, manpower, and resources.

While one unit may use AMPS and a trackball to point and click a digital map to get grid coordinates, a nondigitized unit must still manually plot grids with a protractor. An aircraft that does not have a data transfer device requires the crew to *"fat finger"* in information such as navigation waypoints, which increases mission launch time and fuel burn during runup.

5. The availability of digital map terrain data bases may limit use of the systems.

The Defense Mapping Agency (DMA) produces digital map products on a priority set by Department of Defense (DOD). While most of Southwest Asia maps may be available in digital form, the small island country in one of your contingency plans may not. This is not an aviation–unique problem. DMA is updating its library of digital maps based on the proliferation of digital systems emerging on the battlefield.

6. GPS coverage and accuracy due to jamming or spoofing.

While GPS may be the best system ever invented for navigation, relying solely on GPS without a backup form of navigation (conventional navigation aid (NAVAID), time distance, and heading computations) is not prudent. GPS signals and accuracy depend on factors such as the number of satellites being tracked, constellation orbit, or coding of the signal.

7. Automation skills, such as typing and basic computer skills, as well as operational knowledge, are critical.

Use of digitized systems requires basic skills such as typing. Faster typing means faster data input. Basic knowledge of computer operations such as file management and system operations is essential.

8. The tactical internet may be limited in the volume of data that can be passed.

If too much data is passed, the net could collapse.

Section II. DIGITAL SYSTEMS AND EQUIPMENT

D-8. DESCRIPTIONS OF CURRENT DIGITAL SYSTEMS

This section contains brief descriptions of the current aviation digital systems. For detailed descriptions and technical data, refer to the appropriate system operator's manual.

a. ***AH-64D Longbow Apache (LBA).***

(1) The Longbow system consists of the AH-64D aircraft; provisions for an integrated milimeter wave (MMW) fire control radar (FCR), mounted on top of the Apache's main rotor mast; and the RF Hellfire missile. Not all AH-64Ds will have the MMW radar attached; however, all will be configured to accept the radar. In addition, the AH-64D contains the radio frequency interferometer (RFI) for precise threat system acquisition. The Longbow Apache is able to detect, classify, prioritize, and engage targets with Longbow Hellfire missiles without visually acquiring the target.

(2) The LBA will provide increased data transfer capabilities such as SPOTREPs, SITREPs, battle damage reports, target handover, and real–time images of the battlefield with use of the IDM. The IDM is a tri-service device that offers backward capability to the OH–58D airborne target hand-over system (ATHS) and tactical fire (TACFIRE).

(3) With TACFIRE integration, the Longbow can send target information to the entire TACFIRE net for immediate suppression. Besides target handover, the IDM provides the capability to send fragmentary orders (FRAGOs); enemy/friendly location updates; graphic updates and changes; and any other pertinent information from the commander to the Longbow Apache company/team-or from LBA to commander—while en route to or in the battle position. This timely exchange of information reduces the time of manually plotting information and increases operational tempo, allowing the commander to shape the battlefield.

(4) The data transfer module (DTM), mounted in the Longbow Apache, is used to quickly upload the mission data and initialize aircraft systems for the mission. Data can be loaded into all aircraft or loaded into the first aircraft and transmitted via the IDM to all other aircraft in the mission. When the mission is complete, the DTM retains all mission data debriefing information to carry back to the AMPS for the debrief. In addition, the DTM provides the maintenance section with data for troubleshooting, reducing maintenance downtime.

b. ***OH-58D(I) Enhanced Kiowa Warrior.***

The Kiowa Warrior is a digitized platform. In addition to the digital systems integrated into all OHW–58D(I)s, the Kiowa Warriors taking part in the TF XXI advanced warfighting experiment (AWE) will have the following digitized systems installed:

(1) ***Embedded GPS in an inertial navigation system (EGI).*** The EGI replaces the current doppler/attitude heading reference system (AHRS) combination; it provides increased navigation accuracy.

(2) ***Improved master controller processor unit (IMCPU).*** The IMCPU provides a new digital map display on the multifunction display (MFD). The IMCPU also provides the processing power and memory necessary for integration of the other aircraft upgrades.

(3) ***Improved data modem (IDM).*** The IDM is the Kiowa Warrior's link to the digital battlefield. It replaces the ATHS of the existing Kiowa Warrior; it allows digital data to be transferred over the frequency modulated (FM)I, FM2, ultra–high frequency (UHF), and very high frequency–amplitude modulated (VHF–AM) radios using either TACFIRE or variable message format (VMF) messages.

(4) ***SINCGARS system improvement program (SIP) radio.*** The SINCGGARS SIP replaces the FMI and FM2 radios and the associated data rate adapters (DRAs) in the current Kiowa Warrior. The SINCGARS SIP radio provides faster data communication in a jamming or high–noise environment.

(5) ***Video image crosslink (VIXL).*** The VIXL provides the Kiowa Warrior with the capability to send and receive still frame images over one of the FM radios. The VIXL consists of a circuit card installed in the IMCPU. VIXL ground stations will consist of an AMPS with a tactical communication interface module (TCIM) and a SINCGARS radio. The ground stations will be used to transfer VIXL images on the ground.

(6) ***Improved mast-mounted-sight system processor (IMSP).*** The IMSP is a direct replacement for the existing mast–mounted–sight system processor (MSP) in the Kiowa Warrior. The IMSP provides enhanced targeting through—

- Improved tracking lock-on and reacquisition.
- Television (TV)/Thermal image sensor (TIS) split screen; i.e., provides TV and TIS images on the same display page.
- Auto cue-detects and highlights moving targets.

c. ***Aviation Mission Planning System (AMPS).***

(1) AMPS is an automated aviation mission planning/rehearsal/synchronization tool designed specifically for the aviation commander. The two levels of AMPS are brigade/battalion and company. Each level provides the automated capability to plan, rehearse, and synchronize aviation missions.

(2) The brigade/battalion AMPS is hosted on the common hardware/software II (CHS II) platform. The platform consists of a TCU with 128 megabytes (MB) of random access memory (RAM); a 4.2–gigabytes (GB), removable, hard-disk drive; a compact disk–read only memory (CD–ROM) drive; a 1.3–GB, magneto, optical drive; a 19–inch color monitor; and a character graphics printer. The company AMPS hardware also will be hosted on a CHS II or portable, lightweight computer unit (LCU). All of these components are ruggedized for field use.

(3) In addition, the AMPS has an internal 9600 baud modem. AMPS software contains a modem applet, allowing two AMPS to transfer data files over telephone lines. (A secure STU–III phone must be used if transferring classified data.) Longbow Apache and OH–58D Kiowa Warrior AMPSs have a data transfer receptacle and data cartridge for loading/downloading mission data in the aircraft.

(4) The functions of AMPS can be broken into three areas: tactical planning, mission management, and maintenance management functions. The tactical planning function includes planning tasks normally performed at the brigade/battalion level, such as intelligence data processing; route planning; communication planning; navigation planning; and mission briefing/rehearsal.

(5) The mission management function can be associated with planning that occurs at the company/aircrew level. These tasks include aircraft performance planning; weight and balance calculations; flight planning; crew endurance planning; and OPORD/OPLAN changes. The company system will also be capable of mission briefing/rehearsal.

(6) The maintenance management function if provided primarily for the unit–level maintenance section. This section will permit the downloading of mission data to maintenance personnel.

(7) AMPS data is saved onto a data transfer cartridge (currently 256 kilobytes (KB)), which is used to upload mission data to the host aircraft (AH-64D and OH-58D Kiowa Warrior) via the data transfer module. (Note: the Force XXI Kiowa Warrior will use a 40–MB cartridge to hold digital map data bases.) The data created at battalion level is given to the company level for detailed company and aircrew planning. Integrated into IDM functions, the AMPS will allow information and graphic updates between aircraft, the A^2C^2S, and the AVTOC. Printed hardcopy output products will include weight and balance forms, strip maps, flight plans, OPORDs, route navigation cards, and communications cards.

(8) AMPS map data bases are created from arc digitized raster graphic (ADRG) CD–ROM and digital terrain elevation data (DTED) media available from the DMA. The maps contained on the CD–ROMs are digitally cut and pasted for a particular area of operation; they are stored for ready access on the magneto optical drive disks or the AMPS hard drive. Data bases of different areas of operation or various scale maps can be maintained and organized on disks.

(9) AMPS can be used for detailed terrain analysis; for example, intervisibility line of sight between a battle position and an engagement area. Using the perspective view feature, pilots can gain a feel for prominent terrain along the route to be flown.

(10) AMPS is a standard International Business Machine (IBM)-compatible system. AMPS software uses the UNIX operating system with X–windows environment and a Motif graphical user interface. AMPS has many ports on the back panel that will accommodate commercial, off-the-shelf computer peripherals such as laser printers, mouses, external monitors, or scanners. Some units use the AMPS for more than a mission planner. By using additional removable hard drive and system setups, many units use the AMPS as a reconfigurable tactical work station for word processing, graphics, and data communications. Some units have even obtained commercial video projection devices to project the AMPS display in large–screen format for tactical briefings.

(11) It is important to note that the AMPS in use today is not the objective AMPS. The AMPS currently fielded does not have all the capabilities mentioned because of current hardware and software limitations.

d. ***Army Airborne Command and Control System (A^2C^2S).***

(1) The A^2C^2S is a UH–60 Black Hawk with a console of common networked computers, combat net radios (CNRs), and HaveQuick UHF radios. Satellite communications (SATCOM), HF radios, and a large digital map display on a flat panel screen. This system is used by commanders as a highly mobile C^2 platform.

(2) The A^2C^2S will provide real–time situational awareness and mission planning capability. The A^2C^2S requirement is an extension of the current airborne C^2 capability; it is the airborne variant of the command and control vehicle (C^2V) program. The helicopter

(UH–60)–based C^2 system will provide the commander an airborne C^2 capability—with voice and data equipment. The capability provides battlefield information processing and connectivity equivalent to the tactical command post and the battle command vehicle (BCV) while static or airborne.

(3) Staffing in the A^2C^2S will vary with mission requirements. The system will provide the corps, division, ground maneuver brigades, and attack helicopter battalion commanders with a mobile air vehicle; the vehicle will possess sufficient capability to acquire and communicate critical information at all times. The system will provide the commander an immediate, mobile C^2 node for early entry operations. The airborne C^2 system must fully interoperate with joint and allied forces; the ground commander's combat vehicle (CV); BCVs; corps and below components of the Army tactical command and control system (ATCCS)/Army battle command system (ABCS); and special operation forces C^2 systems.

(4) The A^2C^2S has the capability to communicate and exchange information and graphics with all elements of the battle, such as Longbow Apaches, en route to or in the battle position (BP); ground commander and elements; overhead intelligence sources (joint surveillance target attack radar system (JSTARS); unmanned aerial vehicle (UAV)); fire support element; close air support (CAS); and any other element with the same equipment.

(5) Finally, in peace and humanitarian operations, the system will provide connectivity with civil and/or host nation information/communication networks.

e. ***Aviation Tactical Operations Center (AVTOC).***

(1) The AVTOC is a high–mobility, multipurpose wheeled vehicle with a standard, integrated command post rigid wall shelter, and high–mobility trailer. Aviation brigades, battalions, and separate companies use it to plan and control their forces on the digitized battlefield. It is an integrated system wherein the maneuver control system, aviation mission planner—brigade and below—C^2 exchange data files are connected to a suite of radios and modems. It receives, interprets, parses, and correlates messages from combined arms and joint sources. The AVTOC consists of operations work stations; intelligence work stations; mission planning work stations; reconfigurable work stations, with situational displays; and a large–screen projection system. The communications rack in the vehicle contain VHF AM; UHF AM; VHF FM SINCGARS/SIP; UHF AM HQ II; SATCOM; and HF NOE COMM radios.

(2) In the AVTOC, the aviation commander and his staff coordinate and execute the combat operations in progress; plan future operations; conduct debriefs of crews; and prepare reports for higher headquarters. The division plans and orders will come to the aviation brigade through common computers and nodes to the AVTOC.

f. ***Improved Data Modem (IDM).***

(1) The IDM is used on the A^2C^2S, the AH-64D, the OH-58D Kiowa Warrior, and in the AVTOC. The IDM is a modem that passes targeting or situational awareness information to and from airborne or ground platforms (digital and analog). The IDM contains two modems—which support four links—and one generic interface processor used for link/message processing. (Link formats include TACFIRE, AFAPD, and MTS.)

(2) The IDM provides a demonstrated interoperable capability between the US Air Force, Army, and Marines in pursuit of the digitization of the battlefield. The IDM provides digital connectivity that was previously not available with a low-cost, small–size, and weight design. The IDM can operate simultaneously analog (CPFSK, FSK), digital (ASK), and secure digital (KY-58). The IDM's flexible design is easily adaptable to enable net link and message formats. It is further hardware and software expandable.

g. ***Maneuver Control System (MCS).***

(1) Maneuver control system allows the tactical commander to build a common picture of the battlefield overlayed on DMA digital maps. It has the capability to synchronize battle plans and planning based on near real–time information and assessments from staff and subordinate commanders.

(2) With appropriate communications support, the system displays information and provides the capability to interactively develop the battlefield picture. As a result, the commander is able to make timely decisions and communicate those decisions in graphical formats to higher, adjacent, and lower units. MCS provides the commander and staff the necessary tools for planning, operations order development, and the execution of the battle.

h. ***All Source Analysis System (ASAS).***

(1) ASAS is the IEW component of the Army tactical command and control system (ATCCS) from EAC to battalion. It is a computer–based intelligence and electronic warfare processing, analysis, reporting, and technical control system. ASAS receives, and rapidly processes, large volumes of combat information from multiple sources to formulate timely and accurate targeting information, intelligence products, and threat alerts. It consists of evolutionary sets of hardware and software modules. ASAS provides system operations management; system security; collection management; intelligence processing and reporting; high–value/high-payoff target processing and nominations; and communications processing and interfacing.

(2) ASAS automation provides the G2/S2 and the analysis and control organization of the echelon the means to complete the many analysis tasks required by the IEW functional area as well as those doctrinal missions, tactics, processes, and procedures required by users in various MI organizations. One of these functions is the maintenance of force level data for use by all BFA. This is accomplished at the collateral security classification level by the ASAS Remote Workstation (RWS). The ASAS RWS is manned and operated by the G2/S2 personnel of the echelon; it depends on the products of the highly automated ASAS subsystem supporting the echelon analysis and control element (ACE). All parts of the ASAS are under the OPCON of the G2/S2 of the echelon.

(3) At corps and division, two ASAS RWSs are located at the main command posts. They provide automated intelligence support to the G2 plans and operations staff elements; they also operate as the technical control portion of the IEW C^2 node of the ATCCS. They provide current IEW and enemy situation information to the force level control (FLC) data base for access and use by other ATCCS/FLC subscribers.

(4) At brigade and below, two RWSs are provided to the analysis and control team of the MI direct support company supporting the divisional maneuver brigade. In addition, one RWS is provided to each brigade and battalion S2. These RWSs operate at the collateral security level; they perform the same relative IEW/S2 and ATCCS functions as the division ACE and G2 RWS.

i. ***Enhanced Position Location Reporting System (EPLRS).***

(1) EPLRS is a radio system that provides a robust, transparent, digital communications network with features such as automatic data relay and automatic data reroute. Through time domain multiple access (TDMA) architecture, the EPLRS network is able to accommodate the varied data throughout requirements of multiple users on the battlefield. Through automatic position updates, an individual user of the EPLRS network can be informed of the location of other EPLRS network users. The A^2C^2S will be the only Army aviation aerial platform equipped with EPLRS, although the AVTOC will have it on the ground.

(2) Through the EPLRS established network, the system serves as a situational awareness terminal and brigade data processor. It can graphically display the position of other net users as well as their air tracks (locations). As a brigade data processor, it allows the user to send data messages to other users of the network. Together, EPLRS and situational awareness host software provide a reliable means of acquiring situational awareness information and exchanging data messages.

j. ***Appliqué.***

(1) Appliqué is a system designed to give nondigital systems—air and ground—a strap-on system to digitally interface with other fielded digital systems. It is the primary communications device for the tactical internet. An Appliqué typically will consist of a communications router, a SINCGARS SIP radio, and application software. The Applique will be system dependent. The software uses a windowing system based on X–Windows and a graphical user interface (GUI) that is Motif compatible.

(2) The Appliqué software covers five functional areas:

- Situational awareness.
- Communications management.
- Execution of battle command.
- Supporting battle command.
- Interfaces to battle command.

(3) The situational awareness software provides highly accurate, near real–time situational data and enhanced graphic/visual presentations to provide real–time awareness of the changing situation.

(4) The communications management software will provide a highly flexible, dynamic networking capability that addresses and routes messages with little or no user intervention. It also monitors all communications networks connected to the appliqué.

(5) The battle command execution software will provide the capability to conduct battle command with minimal user interface. It allows rapid creation of maneuver, intelligence, CSS, obstacle, and fire support overlays. It also transmits and receives overlays and messages.

(6) The battle command support software will provide the user with CS and CSS capabilities and the ability to extract information from higher commands' messages.

(7) Finally, Appliqué software will have interfaces with sensor systems that allow the transmission, reception, and display of sensor information. These interfaces will include—

- Dismounted soldier system (DSS).
- Imagery.
- Battle combat identification system (BCIS).
- Intravehicular information system (IVIS).
- All source analysis system (ASAS).
- Advanced field artillery tactical data system (AFATDS).
- Maneuver control system (MCS).

- Forward area air defense command, control, communications, and intelligence system (FAADC³I).
- Combat service support control system (CSSCS).

k. ***Advanced Field Artillery Tactical Data System (AFATDS).***

(1) The US Army Field Artillery School developed the advanced field artillery tactical data system (AFATDS) to control fire support operations. The corps or division artillery brigade formulates the fire support plan; it disseminates this plan through the MCS computer via the area common user system (ACUS).

(2) These communications are mostly digital, using modems and the message formats of AFATDS or the VMF.

l. ***Forward Area Air Defense Command Control, Communications, and Intelligence (FAADC³I) System.***

(1) The US Army Air Defense Artillery School has developed the FAADC³I system. The FAADC³I provides an overlay of battery coverage areas to the MCS computer via ACUS. The FAADC³I system takes information from Air Force, Navy, national, and organic systems to formulate an air picture.

(2) This air picture is too short lived to be of any use in the planning function. It is not, therefore, transmitted on the MCS computer and communications network. Instead, it will be transmitted over EPLRS throughout the FAADC³I network and retransmitted within the local area users network via EPLRS.

m. ***Combat Service Support Control System (CSSCS).***

(1) The US Army Combined Arms Support Command has developed the combat service support control system (CSSCS). The CSSCS is a computerized system for the control of most classes of supplies and equipment and personnel replacements.

(2) It interfaces with the standard Army management information system (STAMIS), which is a series of software programs, such as standard installation division personnel system (SIDPERS), financial systems, etc.

n. ***HaveQuick II (HQ II).***

(1) HaveQuick II is a ground or airborne UHF radio system that is modified to incorporate an electronic counter–countermeasures (ECCM) capability.

(2) A timing signal from an external source maintains synchronization for frequency hopping communications.

o. ***High-Frequency Nap-of-the-Earth Communications (HF NOE COMM).***

High–frequency radios will provide long–range, non–line-of-sight secure and non–secure communications between Army aircraft, AVTOCs, and ATS facilities. They also will provide digital transmission of GPS data to improve the commander's situational awareness.

SECTION III. DIGITALLY EQUIPPED ORGANIZATIONS

The Aviation Restructure Initiative serves as the organizational foundation for all digitally equipped aviation units.

APPENDIX E

ENVIRONMENTAL CONCERNS AND COMPLIANCE

This appendix is a guide by which to attain a balance between accomplishing the mission and protecting environmentally sensitive areas.

E-1. ENVIRONMENTAL AWARENESS

Unit preparations to conduct aviation operations, in any environment, can incorporate the necessary environmental awareness with minimal additional planning. Many aspects of environmental protection discussed below are common sense; most likely, they will be a part of a unit's operational activity.

E-2. PREPARATION FOR TRAINING

Advanced preparation is key to successful completion of training. The same holds true for environmental awareness and protection. The commander should be aware of the publications governing environmental protection. All unit staffs (company and above) should designate an environmental compliance officer/noncommissioned officer (NCO) to serve as unit point of contact (POC). This person will be responsible for environmental education, standing operating procedure (SOP) updates, incident reporting, etc. Army Regulation (AR) 200–1 and AR 200–2 explain the Army's environmental programs. Appendix A, in both regulations, references the additional documents that should be reviewed. Training Circular (TC) 5–400 provides a comprehensive listing of all items of interest to prepare for operating near, and avoiding, environmentally sensitive areas.

a. The following general matrix on POCs should assist in planning for environmental factors affecting unit training:

Topic	Point of Contact
Air pollution	Environmental Division (ED)
Archeological and historic sites	ED and Natural Resources Branch (NRB)
Clean and safe water	ED
Hazardous materials and waste	Directorate of Logistics (DOL), Defense Reutilization and Marketing Office, ED, fire department
Noise pollution	ED, Range Control (Directorate of Plans, Training, and Mobilization (DPTM))
Range clearances and restrictions	Range Control (DPTM)

Topic	Point of Contact
Standing operating procedures	ED
Spill reporting	ED
Threatened/endangered species	NRB
Water pollution	ED
Wetland protection	NRB, Range Control
Wildlife management	NRB, Range Control

b. Most topics can be reviewed by contacting the ED, NRB, and/or Range Control. In most cases, the ED and the NRB are located under the Directorate of Public Works (DPW). In cases in which training is conducted overseas, refer to the host nation equivalent of the POCs listed above. If there is no host nation equivalent, all training will be conducted under US policies and requirements. Units should coordinate with these organizations to provide a briefing before the start of mission training.

E-3. ARMY ENVIRONMENTAL COMPLIANCE ACHIEVEMENT PROGRAM (ECAP)

Units that handle hazardous waste (HW) and hazardous materials (HMs) must designate, in writing, a hazardous waste coordinator. The unit must comply with ECAP protocol; it will be periodically inspected. Obtain ECAP protocols from the Environmental and Natural Resources Division (ENRD)/DPW). If unable to do so, call the Army Environmental Hotline at 1–800–USA–3845 or DSN 584–1699.

E-4. UNIT LEVEL ENVIRONMENTAL PROGRAMS (REFER TO TC 5-400)

Step	Action
1	Ensure all unit personnel have had, or are scheduled to receive, environmental awareness training.
2	Designate, in writing, an environmental compliance officer and an HW coordinator; ensure these persons are properly trained and qualified.
3	The unit environmental compliance officer will interface with appropriate environmental personnel and ensure that the unit complies with environmental laws and regulations.

Step	Action
4	Meet with battalion S3, S4, and installation personnel who deal with environmental issues.
5	Identify requirements concerning ECAP inspections that may affect a unit; identify problem areas and how to avoid them.
6	Ensure SOP addresses environmental issues/procedures that apply to the unit; coordinate environmental requirements with appropriate installation/chain of command personnel.

NOTE: Units should call personnel in the chain of command, DPW, Staff Judge Advocate (SJA), and Range Control for support.

E-5. TYPES OF UNIT PROGRAMS (REFER TO TC 5-400)

a. HM programs.

b. HW programs.

c. Hazardous communications (HAZCOM) programs.

d. Pollution prevention and hazardous waste minimization (HAZMIN) recycling programs.

e. Spill prevention and response plan programs.

NOTE: TC 5-400 gives specific guidance on environmental protection matters and should be complied with. This appendix is intended to supplement, not replace, TC 5-400.

E-6. CONDUCT OF THE MISSION

Environmental concerns pertaining to a mission could be incorporated into the mission briefing using mission, enemy, terrain, troops, and time available (METT-T) factors. Some factors affecting the briefing should be unit mission, geographical location, and time of the year.

a. ***Mission.***

(1) Identify and assess known environmental risks during planning.

(2) Determine environmental impact on mission execution.

(3) Specify those areas to avoid and minimize the effect on units' schemes of maneuver.

(4) Select alternate training methods or goals.

(5) Provide maps/sketches with detailed areas of environmental concern.

(6) Emphasize the importance that every soldier play an active role in the identification, and timely reporting, of new environmental risk elements.

(7) Rapidly and effectively respond to all petroleum, oils, and lubricants (POL) and/or hazardous waste accidents.

(8) Aircraft transporting hazardous material/waste should select routes that allow for quick access in case the aircraft should have to land unexpectedly.

b. ***Enemy.***

(1) Identify areas of probable environmental contamination that could affect friendly force movement.

(2) Evaluate intelligence reports of enemy equipment/capability and how it would be employed against the environment.

(3) Develop enemy target options to minimize environmental effects.

(4) Maneuver enemy action away from environmentally sensitive areas, when feasible.

c. ***Terrain and Weather.***

(1) Provide recommended paths of movement to avoid environmentally sensitive areas.

(2) Emphasize navigation accuracy and identify well-defined terrain features to assist.

(3) Obtain and analyze predominant and developing weather patterns to diminish possible environmental risks.

c.. ***Troops and Equipment.***

(1) Develop a briefing for all soldiers that highlights and defines the environmental concerns/points of interest.

(2) Provide a detailed and accurate SOP that identifies guidelines to avoid risk areas and does not inhibit mission accomplishment.

(3) Anticipate areas of probable risk and brief troops on how to prevent damage.

(4) Employ practice scenarios that test soldier response to changing environmental risks and promote the decision–making process.

(5) Require accurate and timely reports that pertain to any friendly or enemy environmental concerns.

e. ***Time.***

(1) Maximize planning time and minimize complexity of mission brief.

(2) Practice and develop various mission profiles that emphasize adjusting for changing environmental factors, while maintaining the desired momentum.

APPENDIX F

RISK MANAGEMENT

Commanders are responsible for effectively managing risk. They must—

- *Willingly determine the proper balance that will achieve optimum performance from their command.*
- *Select the best risk-reduction options from those that the staff provides.*
- *Acceptor reject residual risk, based on perceived benefits.*

F-1. COMMAND RESPONSIBILITIES

a. Executive Officer.

The executive officer (XO), as director of the staff, ensures integration of risk management in all aspects of staff planning, directing, coordinating, and controlling to support force protection.

b. Staff Officer. In the risk management process, each staff officer must—

- Recommend appropriate control measures.
- Use risk management to assess his or her functional area.
- Recommend appropriate control measures to reduce or eliminate risk.
- Integrate selected risk control into plans and orders.

c. Troop Leaders. Troop leaders must—

- Review control measures for feasibility.
- Report risk issues beyond their control or authority to their seniors for resolution.
- Recommend changes to improve synchronization of their operations in support of the higher commander's plan.
- Use the risk management process to identify, assess, and control hazards for their mission.

F-2. MISSION RISK MANAGEMENT

a. First Three Steps. During the planning of risk management procedures, include the first three steps of the five-step risk management process. These steps are as follows:

(1) Step 1. Identify the major events that are expected to occur during the operation and the hazards associated with all specified and implied tasks. The staff reviews and expands, as appropriate, the list of hazards and major events during the wargame. This procedure helps to ensure that all significant hazards have been identified, and the staff can determine the appropriate force protection measures.

(2) Step 2. Assess hazards. By assessing hazards and evaluating battlefield-framework synchronization, the staff can—

- Figure out the level of risk associated with a given hazard.
- Decide where and when control measures are appropriate to protect the force.

(3) Step 3. Develop controls, and balance a course of action's (COA's) benefits with its potential risks. The staff must—

- Identify hazards and assess risk.
- Focus on critical events first.
- Eliminate unnecessary risks.
- Reduce the amount of mission–essential and prudent risks by applying controls.
- Develop control options that synchronize the operation that eliminate or reduce risks.
- Recommend options for the commander's decision.

Note: In order of priority, options are to eliminate risks through controls or materiel solutions. Leaders should check for residual effects before carrying out risk-reduction options, visualizing what will happen once they implement the option. Often, reducing one risk can create another which, in turn, could introduce other risks or inhibit the execution of Army operations.

b. Risk-Assessment Matrices. Risk–assessment matrices provide a simple analysis method of subdividing an operation into its major operational events to discover areas where the staff might eliminate or reduce risk. The matrix is nearly always more effective than intuitive methods in identifying the extent of risk.

(1) Unit commanders should use the four–level matrix that results in a mission risk assessment of low, medium, high, or extremely high. Other risk management tools that are used to assess hazards should be converted to the four–level matrix to ensure that the resultant hazard assessment product is compatible with risk decision guidance and risk-acceptance approval authority. Units can use the risk-assessment matrix alone or with other analysis techniques.

(2) When using a risk-assessment matrix, the risk assessor must—

- Review each situation to ensure he has evaluated all significant areas of concern, even if the matrix does not include them.
- Use the matrix to analyze risk and target areas of concern for risk-reducing techniques.
- Review individual areas of concern before recommending options.

c. METT-T Risk Assessment. Another technique the risk assessor can use is the mission, enemy, terrain, troops, and time available (METT-T) risk-assessment procedure, Leaders can subjectively decide the likelihood and extent of accidental loss based on this type of analysis. When using the METT–T format, the risk assessor must—

- Determine the mission's complexity and difficulty.
- Assess the enemy situation and identify specific hazards.
- Consider all aspects of the terrain as well as weather and visibility.
- Determine the supervision required and evaluate the experience, training, morale, and endurance of units and their equipment.
- Determine the time available for planning and executing the mission.

d. Fratricide Countermeasures. The commander's decision and supervision of fratricide countermeasures occur later in tactical decisionmaking (after completing the COA analysis). These important points are the means by which the commander benefits from his staff's work. Steps 4 and 5 are included here to preclude oversight:

(1) Step 4. Decide, implement controls, and integrate specific controls into plans, operation orders (OPORDs), standing operating procedures (SOPs), and rehearsals. Knowledge of controls, from the commander to the individual soldier, is essential for successfully implementing and executing controls.

(2) Step 5. Supervise. The commander must enforce controls. Leaders monitor, follow-up, verify, and correct or modify, as appropriate, controls that the commander imposes on his subordinates. When monitoring operational activities, leaders must—

- Avoid administrative intrusions on their subordinates' time.
- Go where the risks are and spend time at the heart of the action.
- Analyze and think through issues, not just watch.
- Work with key personnel to improve operational procedures after the action. (Leaders must not hesitate to assess imminent danger issues on the spot.)
- Fix systemic problems that are hindering combat effectiveness.
- Capture and distribute lessons learned from mishaps and near misses for future use.

F-3. RISK MANAGEMENT RULES

Leaders also must balance the cost of risks with the value of the desired outcome. They must consider and manage risks in making such decisions using the following four general rules:

a. Never accept an unnecessary risk. The leader who has the authority to accept or reject a risk is responsible for protecting his soldiers from unnecessary risks. If he can eliminate or reduce a risk and still accomplish the mission, the risk is unnecessary.

b. Make risk decisions at the appropriate level. The leader who must answer for an accident is the person who should make the decision to accept or reject the risk. In most cases, he will be a senior officer, but small–unit commanders and first-line leaders might also have to make risk decisions during combat. Therefore, they should learn to make. risk decisions during training.

c. Ensure that the benefits of a prudent risk outweigh the possible cost of the risk. Leaders must understand the possible risk and have a clear picture of the benefits to be gained from taking that risk.

d. Integrate protection into planning. Leaders must bring protection into the planning process. If there are means available to protect soldiers from known and possible hazards of a given mission, leaders need to plan for and use them.

APPENDIX G

HISTORY OF ARMY AVIATION

Modern Army aviation was born on 6 June 1942, a few months after the United States entered World War II. Throughout the war, and for several years afterwards, Army aviation was called organic Army aviation. This was done to distinguish it from the Army Air Force and because its aircraft and personnel were organic to battalions, brigades, and divisions of the Army Ground Forces.

The original function of organic Army aviation during World War II was to assist in the adjustment of artillery fire. During the course of the war, however, organic aviation's small fixed-wing aircraft, commonly known as Grasshoppers, came to be used for command and control (C^2), medical evacuation (MEDEVAC), wire laying, courier service, aerial photography, reconnaissance, and other purposes. The principal reason for the expanding mission of organic Army aviation was that its aircraft were accessible to ground commanders and able to operate in close coordination with ground forces. The aircraft of the Army Air Forces often were not.

Both the original creation of organic Army aviation and its assumption of additional functions during World War II provoked friction and rivalry between the Army Ground Forces and the Army Air Forces. When the Army Air Forces became the US Air Force in 1947 and organic Army aviation remained part of the Army, the friction continued and lasted until the 1970s. To avoid the expense of having two aviation organizations with overlapping functions, the War Department and later the Department of Defense (DOD) established restrictions on the roles and missions of Army aviation and on the size and type of Army aircraft.

For essentially the same reason, Army aviation's primary training and the development and procurement of its aircraft were controlled by the Army Air Forces/US Air Force for many years. These restrictions were specified in a series of War Department and DOD memoranda and by agreements between the Army and the Air Force that began in 1942 and continued until 1975. Notwithstanding the continuing restrictions on the roles and missions of Army aviation, its actual functions in combat situations continued to expand during the conflicts in Korea and Vietnam, for essentially the same reasons as during World War II.

Concurrently, Army aviation progressively became independent of the Air Force in matters of training, procurement, and logistics. Army aviation thereby evolved from a small organization with a limited combat support mission to become the principal air arm of the Army. Its expanding mission and responsibilities were reflected in the successive memoranda and agreements—usually negotiated after the exigencies of combat or extensive testing had clearly demonstrated that Army aviation was the logical provider of most of the Army's tactical aerial requirements.

Although Army aviation has continued to use some fixed-wing aircraft up to the present, its evolution to its current role and status resulted to a large degree from the development of the helicopter and of rotary-wing tactics and doctrine. While the Army Air Forces, the Navy, and the Coast Guard acquired helicopters during World War II, Army aviation did not acquire its first one until 1947. The helicopter was in its infancy during that period, however, and, aside from a very few rescue missions, was used only for testing, experimentation, and training. The Army Ground Forces, to which Army aviation was then attached, borrowed a helicopter from the Army Air Forces and conducted tests at Camp Mackall, North Carolina, beginning in 1944; however, no requirement for Army Ground Forces helicopters was established by those tests.

During the early years of the Cold War, the Army Air Forces/US Air Force gave greater emphasis than ever to strategic air operations and correspondingly less emphasis to tactical air support of the Army. The Air Force continued using the helicopter almost exclusively for search–and–rescue operations, reluctant to allocate resources even for testing helicopters for other purposes. The Navy and Coast Guard also continued to use rotary-wing aircraft only for rescue and other similar purposes. Both the Army and the Marine Corps, however, became interested in acquiring helicopters for other uses-especially in view of the growing Army perception that the Air Force had very little interest in tactical transport and close air support (CAS).

In 1946, the War Department Equipment Board determined that Army Ground Forces required four types of helicopters. The types ranged from light liaison to transport helicopters capable of carrying one to three tons and convertible to cargo, passenger, or ambulance use. Three years later, another Army board study expanded these requirements to six types with cargo capacities of up to 25 tons.

Because of the shortage of helicopters and the reluctance of the US Air Force to purchase them for the ground forces, the Army did not acquire its first helicopter, an experimental model of the two-place H–13 Sioux, until 1947. Following initial testing of this aircraft, the Army requested authorization to purchase 150 more and recommended the early development and acquisition of cargo helicopters. The Army was able to acquire a total of only 74 H–13 observation helicopters during the following three years, however, and did not acquire its first cargo/utility helicopter until 1952.

Helicopters were in short supply during the early years of the Korean conflict. In addition to this fact, the Air Force was slow about testing them and resisted procuring them for the Army. When Major General James M. Gavin requested helicopters from the director of requirements for the Air Force, he was told that *"the helicopter is aerodynamically unsound.. and no matter what the Army says, I know that it does not need any."*

During the Korean conflict, the Army used fixed-wing aircraft for essentially the same functions as during World War II. More importantly, however, the war in Korea clearly demonstrated the potential of the helicopter, especially for MEDEVAC and tactical transportation. Although the Army was not able or prepared to employ helicopters for other missions during that period, the Marine Corps successfully demonstrated the helicopter's value in "vertical envelopment" operations—an early version of air mobility and air assault.

Both during and following the Korean conflict, several Army leaders called for the use of helicopters in new tactical missions. General Gavin published an influential article in April 1954, *"Cavalry, and I Don't Mean Horses."* The article called for the use of helicopters in air cavalry operations to provide the mobility differential that Army cavalry forces had lacked in Korea. Much of the conceptual basis for doctrinal development of the helicopter during the 1950s came from General Gavin's vision of a *"sky cavalry"* unit.

The United States Army Aviation School moved from Camp Sill, Oklahoma, to Camp Rucker, Alabama, in 1954. Camp Rucker became Fort Rucker, and the United States Army Aviation Center (USAAVNC) was established there the following year. In 1956, the Aviation School began mounting weapons on helicopters and developing air cavalry tactics. The Aviation School technically was not in total conformity with DOD restrictions on the Army's use of aircraft. However, the School experimented with the arming of helicopters under the auspices of an Army directive to develop *"highly mobile task forces with an improved ratio of fire power to manpower."*

Colonel Jay D. Vanderpool directed most of these combat development experiments. Doctrinal development for this innovative concept was difficult. When the first doctrinal pamphlet on the new sky cavalry unit was written, Colonel Vanderpool, in his own words, *"plagiarized the last field manual written for horse cavalrymen in 1936."* Progress continued, however. By the end of 1960, the basic objective of the Army's air mobility program was that each division have the capability of moving at least one company of infantry by organic airlift.

Developments during the 1960s were considerably more rapid. During the early months of the Kennedy administration, the Army was reorganized according to the "Reorganization Objectives Army Division" (ROAD). This restructuring provided for approximately double the number of aircraft used in the previous organizational structure. Secretary of Defense Robert S. McNamara was not satisfied with this merely quantitative enhancement of the role of Army aviation. In April 1962, he called upon the Army to take a *"bold new look at land warfare mobility. "* The study was to be *"conducted in an atmosphere divorced from traditional viewpoints and past policies, . . .[and] with a readiness to substitute air mobility systems for traditional ground systems wherever analysis shows the substitution to improve our capabilities or effectiveness."*

Secretary McNamara further recommended several Army leaders to take part in the study. Lieutenant General Hamilton H. Howze was included. As the senior officer, General Howze presided over the "Tactical Mobility Requirements Board," commonly known as the "Howze Board." The board completed its work in August 1962. It concluded, *"adoption by the Army of the airmobile concept is necessary and desirable. In some respects, the transition* **is** *inevitable just as was that from animal mobility to motor."*

The board recommended the creation of five air assault divisions in a 16-division Army force structure. Each of these air mobile divisions was to include an air cavalry squadron and to replace 2,339 of its ground vehicles with 459 aircraft.

In 1963, the 11th Air Assault Division tested the concept at Fort Benning, Georgia, and other places. In 1965, the first air mobile division, the 1st Cavalry Division (Airmobile), was organized and sent to Vietnam. Although the concept of air mobility was developed with a mid–intensity European conflict in mind, it proved to be equally valid for the low–intensity conflict in Southeast Asia.

Two other developments during the 1960s made air mobility technically feasible. They reconciled the Army's expanded use of helicopters with DOD policy and regulations. First, the turbine engine for helicopters was developed. This engine gradually replaced the less-efficient, less–powerful, and less–durable reciprocating engine in Army helicopters. Secondly, successive agreements and memoranda had chipped away at the restrictions on the Army's use of helicopters for over 15 years.

The Johnson–McConnell Agreement of 6 April 1966 authorized the Army to develop and employ rotary-wing aircraft for all intratheater purposes, including troop movement and fire support. In return, Army aviation gave up its larger fixed-wing aircraft and became, more so than ever, a rotary–wing force. From 1966 to the present, the Army has been the recognized American leader in the development of helicopters and helicopter weapons, tactics, and doctrine.

Helicopters were first used to transport Vietnamese troops in January 1962. They played an ever–growing role in the conflict in Southeast Asia from that time until the American withdrawal. With the arrival of the UH–1 (Huey), other turbine-powered aircraft, and two airmobile Army divisions, helicopter warfare became the most important innovation of the conflict. The armed helicopter in the tactical role of fire support to the infantry was developed and perfected.

Armed helicopters became essential for providing direct fire support to units operating outside the range of their direct support artillery. Because of the nature of the enemy and the proven value of the helicopter throughout the war, aviation dominated the development of infantry organization and tactics to combat the enemy's light infantry. The airmobile infantry assault supported by aerial fire remained a dominant tactic until 1971. At that time, operation Lamson 719 again raised the question of helicopter survivability on the modern battlefield.

During the early 1970s, the Army conducted a series of tests in Ansbach, Germany. The tests were to determine the suitability of air cavalry elements and the AH–1G Cobra, in particular, to operate in an antitank role in the European environment. These tests demonstrated that antiarmor helicopter teams, properly employed and trained, could achieve high ratios of armored vehicles destroyed for every missile-firing helicopter lost.

Armed scout and attack helicopters-especially when operating in nap-of-the-earth and nighttime environments-clearly were shown to have the required survivability and to be viable and essential elements of conventional mid– to high–intensity warfare. Thus the way was paved for the development of the modern attack and scout helicopters and the doctrinal principles that would take Army aviation into the next century.

The Army—in cooperation with industry-began developing the AH-56 Cheyenne and the AH-64 Apache during the final years of the Vietnam conflict. The Army's development of specifically designed attack helicopters during the 1970s again raised the question of Army and Air Force aerial missions. By this time, the Air Force was content to permit the Army to continue developing helicopters.

The Army continued to desire fixed-wing CAS from the Air Force. It was, therefore, relatively easy for the two services to agree, in 1975, that the attack helicopter did not perform CAS. Instead, it was an extension of organic firepower, and the Air Force would continue to provide CAS with fixed-wing aircraft. The two services thereby agreed to consider the two types of aircraft as complementary rather than duplicative. Since that time, there have been no serious disagreements over aviation missions and functions between the Army and the Air Force.

Throughout the mid– and late 1970s, there was an increasing need to establish a separate Army aviation branch. This step had been seriously considered as early as the mid–1950s. The idea was revived periodically. The opposition to an aviation branch was based in part on the perception that the Army Air Corps had gone its own way and abandoned the ground forces as it became increasingly independent. It was reasoned that a new Army aviation branch would likely do the same thing.

Some also opposed an aviation branch because it was believed that all combat arms branches required aviation support and could best be assured of this support with the existing system. Yet another major basis for opposition stemmed from the expected loss of large funding sources by other branches if aviation, very expensive by its nature, should become a separate branch. Finally, some prominent aviators opposed the establishment of aviation as a separate branch because it was feared that aviation, as a new but very costly branch without senior leaders, would not be able to compete with other branches for funding resources.

The opposition to an aviation branch gradually subsided. First, during the war in Southeast Asia, Army aviation had adequately demonstrated its essential role in modern warfare. At the same time, it had shown that it could cooperate effectively as a member of the combined arms team and that there was no danger in its leaving the rest of the Army behind and going its own way.

Furthermore, as aviation technology and tactics became more and more complex, it was increasingly difficult for aviation officers to be both aviation soldiers and competitive members of their respective branches of the Army. Once trained as aviators, they needed to fly or command other aviators and not spend half of their time in Field Artillery or

Infantry assignments. They needed their own basic and advanced courses in which aviation tactics would be emphasized rather than mentioned in passing.

Aviation also needed to become a branch so that there would be effective central control over the development of its doctrine and equipment. Senior Army leaders encouraged a series of studies during 1980 and 1981. These studies clearly showed, and convinced most doubters, that Aviation should become a separate branch. The US Army Training and Doctrine Command authorized the creation of the branch, and then the Chief of Staff of the Army (CSA). The effective date of the establishment of the Aviation Branch by the Secretary of the Army was 13 April 1983.

Individual aviator training was consolidated at Fort Rucker in 1973. At that time the training programs at Fort Welters, Texas, and Fort Stewart/Hunter Army Airfield, Georgia, were discontinued. After the creation of the Aviation Branch in 1983, further consolidation of aviation–related activities and training under the auspices of USAAVNC and the Aviation Branch chief occurred. Aviation officer courses were implemented at Fort Rucker in 1984. The US Army Air Traffic Control Activity was transferred from the US Army Information Systems Command to USAAVNC two years later. The Noncommissioned Officer Academy was established at Fort Rucker in 1987. In 1988, USAAVNC assumed command and control as well as resource management responsibilities for the Aviation Logistics School at Fort Eustis, Virginia.

The first Army Aviation Modernization Plan (AAMP) was approved and implemented in 1988. As modified in subsequent revisions, this plan called for gradual reduction in the number of Army aircraft as older models were replaced by modern ones. According to the 1992 version of the AAMP, the aircraft inventory of 7,793 aircraft in 1992 would be reduced to 6,150 in 1999 and 5,900 in 2010. Only six aircraft types would be in the rotary–wing fleet in 2010.

The Aviation Restructure Initiative (ARI) was undertaken to correct the deficiencies in Army of Excellence design for aviation units while retiring old aircraft and reducing logistics requirements and costs. The CSA approved the ARI on 3 February 1993, and implementation began in 1994. All forward-deployed forces were scheduled to complete the restructuring process by 1998 and all other units by 1999.

Army aviation units were involved in all major contingency operations during the 1980s and 1990s. In Operation Urgent Fury, the American invasion of Grenada in October 1983, both the Marine Corps and the Army used helicopters. For the Army, Urgent Fury was the first combat test of the new UH–60 Black Hawks, which were used for assault, MEDEVAC, and transport during the operation. Three Army aviation battalions took part in Operation Urgent Fury.

Another new Army helicopter, the OH–58D Kiowa Warrior, was employed in the Persian Gulf in 1987. The Army armed 15 of these aircraft with Hellfire missiles and stationed them on U.S. Navy ships in the Persian Gulf to protect shipping during the war between Iran and Iraq.

Approximately 160 Army helicopters took part in Operation Just Cause, the American invasion of Panama beginning in December 1989. AH-64 Apaches self-deployed from the United States and engaged in combat for the first time. Other Army aircraft performing attack, assault, transportation, and observation roles in Operation Just Cause included Cobras, Black Hawks, Chinooks, Kiowas, and Hueys. The invasion of Panama employed the largest number of special operations aircraft (65 helicopters and 20 fixed-wing planes) ever employed by United States forces. There was general agreement that special operations air support was the best that had ever been provided.

Although the Apache exhibited some mechanical problems during Just Cause, it performed well as an advanced attack aircraft. Lieutenant General Carl Stiner, commander of the XVIII Airborne Corps, was quoted as observing that it could fire a Hellfire missile *"through a window at five miles away at night."* Operation Just Cause enabled Army aviators to demonstrate in combat that, through the use of the night vision devices with which they had trained, they could *"own the night."*

In the early morning of 17 January 1991, an Army aviator fired the first shot of Operation Desert Storm from an Army helicopter. Within a few minutes, two teams of Apaches totally destroyed two Iraqi radar stations, paving way for the air war over Iraq to be conducted with relative impunity.

During the 100–hour ground war, which began about a month later, Army attack helicopters played their most decisive role ever in combat. Most of the Apache's mechanical problems had been corrected. Whatever doubts remained regarding its durability and combat effectiveness were quickly dispelled. Dozens of aviation units and several hundred helicopters of all types took part in the Gulf War. In addition to attack, helicopters were used for air assault, reconnaissance, transportation, combat search and rescue, and observation.

Helicopters, as well as most other types of equipment, were adversely affected by sand and other environmental conditions; however, methods were devised to control the damage and to maintain a high rate of combat readiness. Operation Desert Storm was the first major military operation conducted on a largely electronic battlefield. Army aviation amply demonstrated its effectiveness in this environment and also proved again that it could own the night by carrying out many of its combat operations during darkness.

Since Desert Storm, Army aviation has taken part in several other operations: Provide Comfort in northern Iraq, Restore/Continue Hope in Somalia, Uphold Democracy in Haiti, and the NATO operation in Bosnia. Over 60 Army aircraft and approximately 1,000 aviation personnel operated in Somalia from 1992 to 1994. Somalia provided Army aviation with important lessons relating to military operations in an urban environment. Army aviation's unique combination of versatility, deployability, and lethality cause it to be an indispensable ingredient of almost any type of contingency operation anywhere in the world.

From the sandy beaches and towns of Haiti to the snows and flooded streams of Bosnia, Army aviation continues to demonstrate its unique capability. This unique capability includes infiltration, reconnaissance, evacuation, and strike missions of unconventional warfare and other–than–war operations.

GLOSSARY

Section I. ABBREVIATIONS AND ACRONYMS

A^2C^2	Army airspace command and control
A^2C^2S	Army airborne command and control system
AAMP	Army aviation modernization plan
AATF	air assault task force
AATFC	air assault task force commander
$ABCB^2$	Army battle command brigade and below
ABCS	Army battle command system
ACE	analysis and control element
acft	aircraft
ACR	armored cavalry regiment
ACR(L)	armored cavalry regiment (light)
ACS	armored cavalry squadron
ACT	air cavalry troop
ACUS	area common user system
AD	air defense
ADA	air defense artillery
ADCOORD	air defense coordinator
ADDS	Army data distribution system
ADRG	arc digitized raster graphic
AEB	aerial exploitation battalion

AFAPD	artillery fired atomic projectile device
AFATDS	advanced field artillery tactical data system
AFSO	aerial fire support officer
AG	adjutant general
AH	attack helicopter
AHB	assault helicopter battalion
AHC	assault helicopter company
AHRS	attitude heading reference system
AL	Alabama
ALO	air liaison officer
AM	amplitude modulated
AMPS	aviation mission planning system
AO	area of operations
AOR	area of responsibility
AR	Army regulation
ARI	aviation restructure initiative
ARL	airborne reconnaissance low
ARS	air reconnaissance squadron
ASAS	all source analysis system
ASAS-W	all source analysis system–warrior
ASB	aviation support battalion
ASCC	Army service component command
ASE	aircraft survivability equipment

ASK	audio shift key
ATACMS	Army tactical missile system
ATAS	air–to-air stinger
ATC	air traffic control
ATCCS	Army tactical command and control system
ATHS	airborne target hand-over system
ATKHB	attack helicopter battalion
ATO	air tasking order
ATO-SPINS	air tasking order–special instructions
ATS	air traffic services
AVIM	aviation intermediate maintenance
AVTOC	aviation tactical operations center
AVUM	aviation unit maintenance
AWE	advanced warfighting experiment
BCDSS	battle command decision support system
BCIS	battlefield combat identification system
BCV	battle command vehicle
BDA	battle damage assessment
BM	battle management
BOS	battlefield operating system
BP	battlefield position
BSA	brigade support area
C	cargo airplane

C^2	command and control
C^2I	command, control, and intelligence
C^2V	command and control vehicle
C^2W	command and control warfare
C^3	command, control, and communications
C^3I	command, control, communications, and intelligence
C^4I	command, control, communications, computers, and intelligence
CAB	command aviation battalion
CAC	command aviation company
CAS	close air support
CASEVAC	casualty evacuation
cbt	combat
CD-ROM	compact disk–read only memory
CH	cargo helicopter
CHS II	common hardware/software II
CINC	commander–in-chief
cmd	command
CNR	combat net radio
COA	course of action
COMINT	communications intelligence
COMM	communications
COMMZ	communications zone
COMPO	composite unit

COMSEC	communications security
CONOPS	continuous operations
CONUS	continental United States
COSCOM	corps support command
CP	command post
CPFSK	continuous phase frequency shift key
CS	combat support
CSA	Chief of Staff, US Army
CSAB	combat support aviation battalion
CSAR	combat search and rescue
CSS	combat service support
CSSCS	combat service support control system
CTOC	corps tactical operations center
cu	cubic
CV	combat vehicle
DA	Department of the Army
D^3A	decide, detect, deliver, assess
DCA	defensive counterair
DCSOPS	Deputy Chief of Staff for Operations and Plans
ea	each
DF	direction finding
DISCOM	division support command
DMA	Defense Mapping Agency

DOCC	deep operations coordination cell
DOD	Department of Defense
DOL	Directorate of Logistics
DPTM	Directorate of Plans, Training, and Mobilization
DPW	Directorate of Public Works
DRA	data rate adapter
DRMO	Defense Reutilization and Marketing Office
DS	direct support
DSS	dismounted soldier system
DST	decision support template
DTED	digital terrain elevation data
DTM	data transfer module
DTOC	division tactical operations center
EA	electronic attack; engagement area
EAC	echelons above corps
ECAP	environmental compliance achievement program
ECCM	electronic counter-countermeasures
ECM	electronic countermeasures
ECO	environmental compliance officer
ED	Environmental Division
EGI	inertial navigation system
EH	electronic helicopter
ELINT	electronic intelligence

ENRD	Environmental and Natural Resources Division
EP	electronic protection
EPA	evasion plan of action
EPLRS	enhanced position location reporting system
ES	electronic support
EW	electronic warfare
FAAD	forward area air defense
FAADC^{2}I	forward area air defense command, control, and intelligence
FAADC^{3}I	forward area air defense command, control, communications, and intelligence
FAADS	forward area air defense system
FAC	forward air controller
FARP	forward arming and refueling point
FAST	fast sealift ship
FCR	fire control radar
FEBA	forward edge of the battle area
FID	foreign internal defense
FLC	force level control
FLIR	forward-looking infrared
FLOT	forward line of own troops
FM	frequency modulated; field manual
FMSP	foreign military sales program
FRAGO	fragmentary order
FSCL	fire support coordination line

FSCOORD	fire support coordinator
FSK	frequency shift key
ft	foot; feet
G1	Assistant Chief of Staff, G1 (Personnel)
G2	Assistant Chief of Staff, G2 (Intelligence)
G3	Assistant Chief of Staff, G3 (Operations and Plans)
G4	Assistant Chief of Staff, G4 (Logistics)
gal	gallon
GB	gigabyte
GPS	global positioning system
GS	general support
GSAB	general support aviation battalion
GS-R	general support–reinforcing
GUI	graphical user interface
HAZCOM	hazardous communications
HAZMIN	hazardous waste minimization
HF	high frequency
HHC	headquarters and headquarters company
HHT	headquarters and headquarters troop
HM	hazardous material
HMMWV	high–mobility, multipurpose wheeled vehicle
HQ	headquarters
HQ II	HaveQuick II

hr	hour
HTF	how to fight
HUMINT	human intelligence
HW	hazardous waste
IDM	improved data modem
IEW	intelligence and electronic warfare
IFF	identification, friend or foe (radar)
IMCPU	improved master controller processor unit
IMETP	international military education and training program
IMINT	imagery intelligence
IMSP	improved mast–mounted–sight system processor
in	inch
IPB	intelligence preparation of the battlefield
IVIS	intravehicular information system
J 4	Logistics Directorate
JAAT	joint air attack team
JCS	Joint Chiefs of Staff
J P	jet petroleum
JSAK	joint second echelon attack
JSEAD	joint suppression of enemy air defense
JSTARS	joint surveillance target attack radar system
JTCB	joint targeting coordination board
JTF	joint task force

KB	kilobyte
kt	knot
LAN	local area network
lb	pound
LBA	Longbow Apache
LCC	land component commander
LCU	light–weight computer unit
LIC	low–intensity conflict
LNO	liaison officer (aviation)
lt	light
LUH	light utility helicopter
LWR	laser warning receiver
LZ	landing zone
MB	megabyte
MCS/P	maneuver control system/Phoenix
MEDEVAC	medical evacuation
METL	mission essential task list
METT-T	mission, enemy, terrain, troops, and time available
MFD	multifunction display
MH	modified helicopter
MI	military intelligence
min	minute
MITT	mobile integrated tactical terminal

MLRS	multiple launch rocket system
MMW	milimeter wave
MOOTW	military operations other than war
MOUT	military operations on urbanized terrain
MOPP	mission–oriented protective posture
MOUT	military operations on urbanized terrain
MSCA	military support to civil authorities
MSP	mast–mounted–sight system processor
MTS	Marine tactical system
NA	not applicable
NAI	named area of interest
NATO	North Atlantic Treaty Organization
NAVAID	navigational aid
NBC	nuclear, biological, chemical
NCA	national command authority
NCO	noncommissioned officer
NEO	noncombatant evacuation operations
nm	nautical mile
NOE	nap–of–the-earth
NRB	Natural Resources Branch
NRT	near real time
OCA	offensive counterair
OH	observation helicopter

OPCON	operational control
OPLAN	operational plan
OPORD	operation order
OPSEC	operations security
OPTEMPO	operational tempo
PIR	priority intelligence requirement
POC	point of contact
POL	petroleum, oils, and lubricants
PSYOP	psychological operations
PZ	pickup zone
RAH	reconnaissance attack helicopter
RAM	random access memory
RAS	regimental aviation squadron
RC	reconnaissance cargo airplane
RF	radio frequency
RFI	radio frequency interferometer
ROAD	reorganization objectives Army division
RPV	remotely piloted vehicle
RU	reconnaissance utility airplane
RV	reconnaissance vehicle
RWS	remote workstation
S1	adjutant (US Army)
S2	intelligence officer (US Army)

S3	operations and training officer (US Army)
S4	supply officer (US Army)
SAC	support aviation company
SAR	search and rescue
SASO	stability and support operations
SATCOM	satellite communications
SBF	support by fire
SEAD	suppression of enemy air defense
SEMA	special electronic mission aircraft
SERE	survival, evasion, resistance, and escape
SHORAD	short range air defense
SIDPERS	Standard Installation Division Personnel System
SIGINT	signals intelligence
SINCGARS	single channel ground and airborne radio system
SIP	system improvement program
SITREP	situation report
SJA	Staff Judge Advocate
SOA	special operations aviation
SOF	special operations forces
SOP	standing operating procedure
SPOTREP	spot report
STAMIS	standard Army management information system
TA	theater Army

TAA	tactical assembly area
TAACOM	theater Army area command
tac	tactical
TACAIR	tactical air
TACFIRE	tactical fire
TACP	tactical air control party
TAI	target area of interest
TBM	theater ballistic missile
TC	training circular
TCAE	technical control and analysis element
TCF	tactical combat force
TCIM	tactical communication interface module
TCU	tactical computer unit
TDMA	time domain multiple access
TENCAP	tactical exploitation of national capabilities
TF	task force
TIS	thermal image sensor
TMD	theater missile defense
TOC	tactical operations center
TOE	table(s) of organization and equipment
TOW	tube launched, optically tracked, wire guided
TRADOC	United States Army Training and Doctrine Command
TV	television

UA	utility airplane
UAV	unmanned aerial vehicle
UH	utility helicopter
UHF	ultra-high frequency
US	United States (of America)
USA	US Army
USAAVNC	US Army Aviation Center
USAF	United States Air Force
UV	utility vehicle
VHF	very high frequency
VIXL	video image crosslink
VMF	variable message format
XO	executive officer

Section II. TERMS AND DEFINITIONS

AC— Active Component

ACR— armored cavalry regiment

ADA— air defense artillery

agility— the ability of friendly forces to act faster than the enemy.

air assault— operations in which air assault forces—using the firepower, mobility, and total integration of helicopter assets in their ground or air roles—maneuver on the battlefield under the control of the ground or air maneuver commander to engage and destroy enemy forces.

air combat— Any use of organic weapons to engage another aircraft while both aircraft are in flight.

air defense—all defensive measures designed to destroy attacking enemy aircraft or missiles in the earth's envelope of atmosphere or to nullify or reduce the effectiveness of such an attack. (NATO)—all measures designed to nullify or reduce the effectiveness of hostile air actions. (Army aviation does not perform counterair operations. However, Army aviation units do participate in the air defense effort by conducting air combat operations.)

air liaison officer—the senior Air Force officer at each TACP who advises the Army commander and staff on the capabilities, limitations, and employment of tactical air operations. He operates the Air Force request net. He coordinates CAS missions with the fire support element and assists it in planning the simultaneous employment of air and surface fires. He supervises FACs and will assist the fire support team in directing airstrikes in the absence of a FAC.

air movement operations—operations using airlift assets—primarily helicopters—to move maneuver, CS, and CSS forces and/or equipment whose primary purpose is not to engage and destroy enemy forces. **(NATO) air movement**—air transport of units, personnel, supplies, and equipment, including airdrops and air landings.

air reconnaissance—the acquisition of intelligence information by employing visual observation/sensors in air vehicles. **(NATO)**— the collection of information of intelligence interest either by visual observation from the air or through the use of airborne sensors.

air tasking order— the air operations center tasking order (USAF) to the air assets under its control. Similar in purpose to the OPORD. Although not under the operational control of the USAF, Army SEMA assets may use the ATO–SPINS (air tasking order–special instructions) to describe their missions to other ATO–SPINS recipients.

antiterrorism— defensive measures to reduce the vulnerability of individuals and property to terrorism.

area defense—denying the enemy access to designated terrain for a specific time to retain ground using a combination of defensive positions and small, mobile reserves.

area of operations—a geographical area assigned to an Army commander by a higher commander. An AO has lateral and rear boundaries that usually define it within a larger joint geographical area.

armed helicopter— a helicopter fitted with weapons or weapon systems.

Army force— that force provided by the Army service component to the subordinate joint force commander for the conduct of joint operations.

Army service component commander— the commander of the Army service component command assigned to the theater combatant commander who provides forces and support to subordinate joint force commanders. The ASCC is responsible for preparing; maintain ing, training; equipping; administering; and supporting Army forces assigned to unified and specified commands. The ASCC normally advises the combatant or subordinate unified commander on the proper employment of the forces of the Army component. The ASCC is normally not a part of the joint force staff.

assign— to place units or personnel in an organization in which such placement is *relatively permanent* and/or in which such an organization controls and administers the units or personnel for the primary function or greater portion of the functions of the unit or personnel.

attach— to place units or personnel in an organization where such placement is *relatively temporary*. Subject to limitations imposed in the attachment order, the commander of the formation, unit, or organization receiving the attachment will exercise the same degree of command and control over them as he does over the units and persons organic to his command. However, the responsibility for transfer and promotion of personnel will normally be retained by the parent formation, unit, or organization.

attack helicopter—a helicopter designed to search out, attack, and destroy enemy targets.

aviation maneuver forces— aviation maneuver units that operate in the ground environment. They engage targets by fire from covered and concealed positions. Their operations are similar to ground maneuver operations in that they tailor their movement to the terrain and use supporting fires. These units are integrated into the tactical plan of the ground force commander. They can control terrain by denying the enemy its use by direct aerial fire for limited periods of time.

battle command—the art of battle decision making, leading, and motivating soldiers and their organizations into action to accomplish missions. Includes visualizing current state and future state, then formulating concepts of operations to get from one to the other at least cost. Also includes assigning missions; prioritizing and allocating resources; selecting the critical time and place to act; and knowing how and when to make adjustments during the fight.

battle damage assessment— the process of determining the essential tactical reconstitution requirements for an attrited unit; the process of determining the combat effectiveness of the enemy after engagement by friendly force.

battlefield framework—an area of geographical and operational responsibility established by the commander; it provides a way to visualize how he will employ his forces; it helps him relate his forces to one another and to the enemy in time, space, and purpose.

battlefield operating systems— the major functions performed by the force on the battlefield to successfully execute Army operations (battles and engagements) to accomplish military objectives directed by the operational commander; they include maneuver, fire support, air defense, command and control, intelligence, mobility and survivability, and combat service support.

battle space— components determined by the maximum capabilities of a unit to acquire and dominate the enemy; includes areas beyond the AO; it varies over time according to how the commander positions his assets.

cargo (transport) helicopter—a helicopter designed and employed primarily for movement of passengers/cargo.

close support— the action of the supporting force against targets or objectives that are sufficiently near the supported force to require detailed integration or coordination of the supporting action with the fire, movement, or other actions of the supported force.

combat maneuver forces— forces that use fire and movement to engage the enemy with direct fire weapon systems, as distinguished from forces that engage the enemy with indirect fires or otherwise provide combat support. These maneuver elements are primarily infantry, armor, cavalry (air and armored), and *aviation.*

combat service support— the focus of logistics at the tactical level of war; the synchronization of essential functions, activities, and tasks necessary to sustain soldiers and their weapon systems in an area of operations; includes but is not limited to that support rendered by service support troops to arm, fuel, fix, move, man, and sustain soldiers and their equipment.

combat support— fire support and operational assistance provided to combat elements. Artillery, air defense artillery, engineer, military police, signal, military intelligence, chemical, and *aviation* all provide CS.

combatting terrorism— actions—including antiterrorism (defensive measures taken to reduce vulnerability to terrorist acts) and counterterrorism (offensive measures taken to prevent, deter, and respond to terrorism)—taken to oppose terrorism throughout the entire threat spectrum.

combined arms— application of several arms, such as infantry, armor, artillery, and *aviation.*

combined arms team— two or more arms mutually supporting one another. A team can consist of armor, infantry, cavalry, *aviation,* field artillery, air defense artillery, and engineers.

commander's intent—a concise expression of the purpose of an operation, a description of the desired end state, and the way in which the posture of that goal facilitates transition to future operations.

communications zone— the rear part of the theater of war or theater of operations that contains the lines of communications, theater logistics bases, forward operating bases, and other agencies required for the immediate support and maintenance of the field forces; extends back to the CONUS base.

conflict— the state characterized by confrontation and the need to engage in hostilities to secure strategic objectives.

counterattack— an attack with a reserve or lightly committed forward element that is launched after the enemy begins its attack, after the commander has identified the enemy's effort, or when a resolute defense creates an assailable enemy flank.

culmination— the point in time and space when the attacker's combat power no longer exceeds that of the defender or when the defender no longer has the capability to defend successfully.

decisive force— applying overwhelming forces to fight and win quickly with minimum casualties.

decisive point—a point, usually geographical in nature, that, when retained, provides a commander with a marked advantage over his opponent. Decisive points could also include other physical elements such as enemy formations, command posts, and communications nodes.

deep operations—operations designed in depth to secure advantages in later engagements; protect the current close fight; and defeat the enemy more rapidly—by denying freedom of action and disrupting, or destroying, the coherence and tempo of its operations.

deliberate attack—fully synchronized operations that employ the effects of every available asset against the enemy's defense.

demobilization— the act of returning the force and materiel to a premobilization posture or to some other approved posture; also involves returning the mobilized portion of the industrial base to peacetime conditions.

demonstration—a show of force in an area where a decision is not sought, made with the aim of deceiving the enemy as to the true intentions of the attack.

deployment— the relocation of forces to desired areas of operations; the movement of forces within areas of operations.

direct support—(DOD)—a mission requiring a force to support another specific force and authorizing it to answer directly the supported force's request for assistance. **(NATO)**— the support provided by a unit or formation not attached or under command of the supported unit or formation but required to give priority to the support required by that unit or formation.

doctrine— fundamental principles by which the military forces, or elements thereof, guide their actions in support of national objectives. It is authoritative but requires judgment in application. **(NATO)**— fundamental principles by which the military forces guide their actions in support of objectives. It is authoritative but requires judgment in application.

electronic warfare—military actions including **a. electronic attack**—the use of either electromagnetic or directed energy to degrade, neutralize, or destroy an enemy's combat capability; **b. electronic protection**— those actions taken to protect personnel, facilities, and equipment from any effects of friendly or enemy employment of electronic warfare; **c. electronic warfare support**— those actions tasked by an operational commander to search for, intercept, identify, and locate sources of radiated electromagnetic energy for the purpose of immediate threat recognition.

engagement— small, tactical conflicts, usually between opposing maneuver forces.

envelopment— an offensive maneuver in which the main attacking force passes around or over the enemy's principal defensive positions to secure objectives in the enemy's rear.

exploitation— the attacker's extension of destruction of the defending force by maintaining offensive pressure.

feint— a spoiling attack designed to divert the enemy's attention from the main effort.

fire support— assistance to those elements of the ground forces that close with the enemy, such as infantry and armor units, rendered by delivering artillery and mortar fire, naval gunfire, and close air support. Fire support may also be provided by tanks, air defense artillery, and armed helicopters.

forward air controller— A member of the TACP who, from a ground or airborne position, controls aircraft engaged in CAS of ground forces.

fratricide— the employment of friendly weapons and munitions with the intent to kill the enemy or destroy his equipment or facilities, which results in unforeseen and unintentional death or injury to friendly personnel.

general support— that support given to the supported force as a whole and not to any particular subdivision thereof.

general support—reinforcing— a tactical artillery mission. GS–R is not used by aviation forces.

hasty attack— result of a meeting engagement—launched with the forces at hand and with minimum preparation to destroy the enemy before he is able to concentrate or establish a defense.

health services— the logistical function of promoting, improving, conserving, or restoring the mental or physical well being of military personnel.

high-payoff target—high–value targets that, if successfully attacked, would contribute substantially to the success of friendly plans.

high-value target—a target whose loss to the enemy can be expected to contribute to substantial degradation of an important battlefield function.

host nation support—civil and/or military assistance rendered by a nation to foreign forces within its territory during peacetime, times of crisis, emergencies, or war; assistance provided during war is based upon agreements mutually concluded between nations.

humanitarian assistance—assistance provided by DOD forces, as directed by appropriate authority, in the aftermath of natural or man-made disasters to help reduce conditions that present a serious threat to life and property; assistance provided by US forces is limited in scope and duration and is designed to supplement efforts of civilian authorities who have primary responsibility for providing such assistance.

identification friend or foe—a system using electromagnetic transmissions to which equipment carried by friendly forces automatically responds: for example, by emitting pulses thereby allowing friendly forces to be distinguished from enemy forces.

initiative—the ability to set or change the terms of battle; implies an offensive spirit.

intelligence— the product resulting from collection, processing, integration, analysis, evaluation, and interpretation of available information concerning foreign countries or areas.

intelligence— preparation-of-the-battlefield-A systematic and continuous process which describes the tactical environment and the effects of that environment on operations and what the enemy can accomplish.

interdiction—actions to divert, disrupt, delay, or destroy the enemy before it can affect friendly forces.

interoperability—The ability of systems, units, or forces to provide services to and accept services from other systems, units, or forces and to use the exchanged services to enable them to operate effectively together.

joint— connotes activities, operations, and organizations in which elements of more than one service of the same nation participate.

joint task force— a force composed of assigned or attached elements of two or more services and constituted by appropriate authority for a specific or limited purpose or missions of short duration.

lines of communication— All the routes (land, water, and air) that connect an operating military force with a base of operations and along which supplies and military forces move.

logistics— the process of planning and executing the movement and sustainment of forces in the execution of military operations. Logistics includes the design, development, acquisition, storage, movement, distribution, maintenance, evacuation, and disposition of materiel; the acquisition, preparation, maintenance, equipping, movement, and health support of personnel; the acquisition or furnishing of services; and the acquisition, construction, maintenance, operation, and disposition of facilities. Logistics is an overarching function that must encompass the range of military operations. At the tactical level, logistics focuses on the traditional CSS functions of arming, fixing, fueling, manning, moving, and sustaining soldiers.

logistics-over-the-shore operations— the loading and unloading of ships without the benefit of fixed port facilities.

maneuver— the movement of forces supported by fire to achieve a position of advantage from which to destroy or threaten destruction of the enemy.

mobile defense— employing a combination of fire and maneuver, offense, defense, and delay to destroy the enemy and defeat his attack.

mobilization— the process by which the Armed Forces, or a portion thereof, is brought to a state of readiness for war or other national emergency; includes activating all or part of the RC, as well as assembling and organizing personnel, supplies, and materiel.

modular units— units comprised of multiple capabilities; depending on the requirement, modules can be added or subtracted from the unit or force package.

multinational operations— An operation conducted by forces of two or more allied nations acting together to accomplish a single mission.

nation assistance— diplomatic, economic, informational, and military cooperation between the US and the government of another nation, with the objective of promoting internal development and the growth of sustainable institutions within that nation. This corrects conditions that cause human suffering and improves the quality of life of the nation's people.

NBC defense— the methods, plans, procedures, and training for defense against the effects of an attack by NBC weapons. The term may not be used in the context of US offensive operations.

noncombatant evacuation operations— operations that relocate threatened civilian noncombatants from locations in a foreign country or host nation. These operations normally involve US citizens whose lives are in danger. They may also include selected host nation natives and third country nationals.

observation helicopter—helicopter used primarily for observation and reconnaissance but which may be used for other roles.

operational control—the authority delegated to a commander to direct forces assigned so that the commander may accomplish specific missions or tasks that are usually limited by function, time, or location; to deploy units concerned and to retain or assign tactical control of those units. It *does not* include authority to assign separate employment of components of the units concerned. Nor does it, of itself, include administrative or logistical control.

operational operating systems— the major functions performed by joint and combined operations forces to successfully execute campaigns and major operations in a theater or area of operations; these systems include movement and maneuver, fires, intelligence, protection, command and control and support.

operations in-depth— the totality of the commander's operations against the enemy—composed of deep, close, and rear operations that are usually conducted simultaneously in a manner that appears as one continuous operation against the enemy.

organic— assigned to and forming an essential part of a military organization; an element normally shown in the unit's table of organization and equipment.

pathfinders—1. Teams dropped or airlanded at an objective to establish and operate navigational aids to guide aircraft to drop zones/landing zones. **2.** Teams air-delivered into enemy territory for determining the best approach and withdraw lanes, landing zones, and sites for heliborne forces.

peace building— postconflict diplomatic and military action to identify and support structures that tend to strengthen and solidify peace to avoid a relapse into combat.

peace enforcement—military intervention to forcefully restore peace between belligerents who may be engaged in combat.

peacekeeping— operations using military forces and/or civilian personnel, at the request of the parties to a dispute, to help supervise a cease-fire agreement and/or separate the parties.

peacemaking— the diplomatic process or military actions to gain an end to disputes.

power projection— the ability of the nation to apply all or some of the instruments of national power—diplomatic, economic, informational, or military—to respond to crisis, to contribute to deterrence, and to enhance regional stability.

priority of support— priorities set by the commander in his concept of the operation and during execution to ensure that combat support and combat service support are provided to subordinate elements in accordance with their relative importance to accomplishing the mission.

pursuit— an attack with the purpose of annihilating the enemy once his resistance has broken down completely and he is fleeing the battlefield.

raid— a limited–objective attack into enemy territory *not* for the specific purpose of gaining and holding ground.

rear operations— operations that assist in providing freedom of action and continuity of operations, logistics, and battle command. Their primary purposes are to sustain the current close and deep fights and to posture the force for further operations.

reconnaissance— a mission undertaken to obtain information by visual observation or other detection methods about the activities and resources of an enemy or potential enemy or about the meteorologic, hydrographic, or geographic characteristics of a particular area.

reconstitution— at the strategic level, those functions and activities required to restore the Army's capability to respond to any mission across the full range of possible operations. At the operational and tactical levels, reconstitution consists of extraordinary actions that commanders plan and implement to restore units to a desired level of combat effectiveness commensurate with mission requirements and available resources.

reinforcing— in artillery usage, a tactical mission in which one artillery unit augments the fire of another artillery unit. Reinforcing is not used by aviation forces.

retrograde operation— a maneuver to the rear or away from the enemy to improve a situation or prevent a worse situation from occurring.

risk management— the process of detecting, assessing, and controlling risk arising from operational factors and making decisions that balance risk costs with mission benefits.

rules of engagement— directives issued by competent military authority that delineate the circumstances and limitations under which US forces will initiate and/or continue combat engagement with other encountered forces.

security assistance— groups of programs authorized by the Foreign Assistance Act of 1961, as amended, and the Arms Export Control Act of 1976, as amended, or other related statutes by which the United States provides defense articles, military training, and other defense–related services, by grant, loan, credit, or cash sales in furtherance of national policies and objectives.

split-based logistics— dividing logistics management functions so that only those functions absolutely necessary are deployed, allowing some management functions to be accomplished from CONUS or another theater.

spoiling attack— an attack from a defensive posture to disrupt an expected enemy attack. A spoiling attack attempts to strike the enemy while he is most vulnerable during his preparations for attack in assembly areas and attack positions—or while he is on the move before crossing the line of departure.

stability and support operations— military activities during peacetime and conflict that do not necessarily involve armed clashes between two organized forces.

strategic mobility— transportation actions using national assets, both military and civilian, in support of a force-projection mission.

support—(DOD)— the action of a force that aids, protects, complements, or sustains another force in accordance with a directive requiring such action. **(NATO)**— The action of a force-or portion thereof—that aids, protects, complements, or sustains any other force.

supporting attack— an attack designed to hold the enemy in position, to conceal the location of the main attack, to prevent him from reinforcing the elements opposing the main effort, and/or to cause him to commit his reserves prematurely at an indecisive location.

supporting fire— fire delivered by supporting units to assist or protect a unit in combat.

supporting forces— forces stationed in, or to be deployed to, an area of operations to provide support for the execution of an operation order. Operational command of supporting forces is not passed to the supported commander.

surveillance— the systematic observation of aerospace, surface or subsurface area, places, persons, or things, by visual, aural, electronic, photographic, or other means.

synchronization— the ability to focus resources and activities in time and space to produce maximum relative combat power at the decisive point.

tactical air control party— advises and assists the commander, requests and coordinates tactical air support, and meets other requirements of the individual force echelon support. The TACP consists of experienced personnel; ground/airborne vehicles; and the communications equipment required to obtain, coordinate, and control tactical air support of ground operations.

tactical combat forces— those forces the commander assigns the mission of defeating enemy airborne, air assault, amphibious, or ground-infiltration attacks in the rear area.

Tactical Exploitation of National Capabilities—a program of the *Army Space Program Office* with the charter: To exploit the current and future tactical potential of nation space systems and integrate the capabilities into the Army's tactical decision making progess as rapidly as possible. National space systems make available near real time imagery intelligence (NRT IMINT) and near real time signals intelligence (NRT SIGINT) derived from national level sources to Army commanders. National space systems at the division echelon include equipment such as the mobile integrated tactical terminal (MITT).

tactics— the art and science of employing available means to win battles and engagements.

tempo— the rate of military action; controlling or altering that rate is a necessary means to initiative; all military operations alternate between action and pauses as opposing forces battle one another and fight friction to mount and execute operations at the time and place of their choosing.

tenets— a basic truth held by an organization; the fundamental tenets of Army operations doctrine describe the characteristics of successful operations.

terrain flight—the tactic of employing helicopters to use the terrain's vegetation and man–made objects to degrade the enemy's ability to visually, optically, or electronically detect or locate the helicopter. This tactic involves a constant awareness of the capabilities and positions of enemy weapons and detection means in relation to available masking terrain features and flight routes. Terrain flying involves flight close to the earth's surface and includes the tactical application of low–level, contour, and nap-of-the-earth flight techniques.

utility helicopter— a multipurpose helicopter capable of lifting troops but which may be used in command and control, logistics, casualty evacuation, or armed helicopter roles.

versatility—the ability of units to meet diverse challenges, shift focus, tailor forces, and move from one role or mission to another rapidly and efficiently.

war— a state of open and declared armed hostile conflict between political units such as states or nations; may be limited or general in nature.

REFERENCES

Army Regulations

AR 200–1. Environmental Protection and Enhancement, 23 April 1990.
AR 200–2. Environmental Effects of&my Actions, 23 December 1988.

Field Manuals

FM 8–10–6. Medical Evacuation in a Theater of Operations, Tactics, Techniques, and Procedures, 31 October 1991.
FM 34–1. Intelligence and Electronic Warfare Operations, 27 September 1994.
FM 34–25–3. All–Source Analysis System and the Analysis and Control Element, 3 October 1995.
FM 34–25–7. (S) Special Electronic Mission Aircraft Survivability (U), 3 October 1995.
FM 1–108. Doctrine for Army Special Operations Aviation Forces, 3 November 1993.
FM 1–111. Aviation Brigades, 27 August 1990.
FM 1–112. Tactics, Techniques, and Procedures for the Attack Helicopter Battalion, 21 February 1991.
FM 1–113. Assault Helicopter Battalion, 28 October 1986.
FM 1–114. Tactics, Techniques, and Procedures for the Regimental Aviation Squadron, 20 February 1991.
FM 1–116. Tactics, Techniques, and Procedures for the Air Cavalry/Reconnaissance Troop, 20 February 1991.
FM 44–100. US Army Air Defense Operations, 15 June 1995.
FM 90–18. Multi–Service Procedures for Combat Search and Rescue, 31 May 1991.
FM 100–5. Operations, 14 June 1993.
FM 100–6. Information Operations, (Draft/TBP)
FM 100–7. Decisive Force: The Army in Theater Operations, 31 May 1995.
FM 100–16. Army Operational Support, 31 May 1995.
FM 100–20. Military Operations in Low Intensity Conflict, 5 December 1990.
FM 100–23. Peace Operations, 30 December 1994.
FM 100–103. Army Airspace Command and Control in a Combat Zone, 7 October 1987.

Training Circulars

TC 1–210. Aircrew Training Program Commander's Guide to Individual and Crew Standardization, 3 October 1995.
TC 5–400. Unit Leader's for Environmental Stewardship, 29 September 1994.

US Army Training and Doctrine Command Pamphlet

TRADOC Pam 525–5, Force XXI Operations, 1 August 1994.

INDEX

This index is organized alphabetically by topics and subtopics identified by page numbers.

Made in the USA
Monee, IL
04 November 2026

67382860R00095